Modern Methods of Reflector Antenna Analysis and Design

The Artech House Antenna Library

Helmut E. Schrank, *Series Editor*

Blake, L.V., *Antennas*

Gupta, K.C., and A. Benalla, eds., *Microstrip Antenna Design*

Hirsch, Herbert L., and Douglas C. Grove, *Practical Simulation of Radar Antennas and Radomes*

Kumar, A., and H.D. Hristov, *Microwave Cavity Antennas*

Law, Preston E., Jr., *Shipboard Antennas*

Li, S.T., J.C. Logan, J.W. Rockway, and D.W.S. Tam, *Microcomputer Tools for Communications Engineering*

Li, S.T., J.C. Logan, J.W. Rockway, and D.W.S. Tam, *The MININEC System: Microcomputer Analysis of Wire Antennas*

Pozar, David M., *Antenna Design Using Personal Computers*

Sletten, Carlyle J., ed., *Reflector and Lens Antennas: Analysis and Design Using Personal Computers*

Sletten, Carlyle J., ed., *Reflector and Lens Antennas: Software User's Manual and Example Book*

Weiner, M.M., et al., *Monopole Elements on Circular Ground Planes*

Wolff, Edward A., *Antenna Analysis: Revised 2nd Edition*

Kitsuregawa, Takashi, *Advanced Technology in Satellite Communication Antennas: Electrical and Mechanical Design*

Scherer, James P., *LAARAN: Linear Antenna Array Analysis Software and User's Manual*

Modern Methods of Reflector Antenna Analysis and Design

Craig Scott

Rockwell International
El Segundo, California

Artech House
Boston • London

Library of Congress Cataloging-in-Publication Data

Scott, Craig.
   Modern methods of reflector antenna analysis and design / Craig
Scott.
      p.   cm.
   Includes bibliographical references .
   ISBN 0-89006-419-9
   1. Antennas, Reflector. I. Title.
TK7871.6.S37 1990   89-9644
621.382'4--dc20   CIP

British Library Cataloguing in Publication Data

Scott, Craig
   Modern methods of reflector antenna analysis and design.
   1. Telecommunication equipment. Reflector antennas.
   Mathematics
   I. Title
   621.3824

   ISBN 0-89006-419-9

Published by

ARTECH HOUSE
685 Canton Street
Norwood, MA 02062

International Standard Book Number: 0-89006-419-9
Library of Congress Catalog Card Number: 89-49644

10   9   8   7   6   5   4   3   2   1

To Tracy and to my parents

Contents

Preface

Sophisticated new integration techniques have been developed in recent years which have revolutionized the way in which radiation patterns are calculated for large, focused reflector antennas. As a result, reflector antenna analysis has become a much more specialized (and, at first glance, more difficult) field within electromagnetics. This book was written to provide a comprehensive treatment of these new methods and to integrate them as elements of contemporary reflector antenna theory.

The integration methods presented in this book include the Ludwig algorithm, Rusch's formula for symmetrical reflectors (and Wong's extension to laterally displaced feeds), the quadratic phase method of Pogorzelski, the sampling methods of Bucci *et al.* and Rahmat-Samii, the Fourier-Bessel approach of Mittra *et al.*, and the Jacobi-Bessel method of Galindo *et al.* Of these, only the last method involves mathematics which might be unfamiliar to most students and engineers. Fortunately, however, the analytical difficulty of this method belies its computational simplicity.

In addition to modern analysis techniques, this book also presents modern methods for the design of large reflector antennas. Modern antenna design involves much more than the old "antenna design slide rule" approach of calculating reflector diameter from the gain requirements and then calculating an F/D ratio on the basis of a given feed pattern beamwidth. This approach (which will also be explained in detail in the text) is the natural starting point for an antenna design, but it is only the beginning. Modern antenna design now involves sophisticated reflector shaping techniques for symmetrical and offset dual reflector systems.

This book will find suitable use as a text for a semester course in reflector antenna theory at the first year graduate level. In addition, it is also easily accessible to practicing antenna engineers. The only prerequisites are a senior level knowledge of general electromagnetic theory and some knowledge of Fourier analysis techniques. Beyond these modest foundations, almost all of the mathematics is carefully derived from basic principles, making the book virtually self-contained. References are cited liberally but are not necessary for understanding the theory. They are included primarily to acknowledge the originators of the theory.

Acknowledgements

I want to thank everyone at Artech House for really making publishing a joy. Also, the reviewers deserve special mention for their many worthwhile comments and suggestions. I am indebted to Ron Pogorzelski for introducing me to this subject and providing me with the opportunity to learn it in some depth. I am also grateful to John Ruze for generously lending invaluable insight on the subject of the Petzval Surface.

Chapter 1
Reflector Antenna Concepts

All of the methods presented in this book for calculating the far-field radiation pattern of a reflector antenna revolve around just a few electromagnetics concepts. Principal among these is the equation for the electric field produced by a current distribution [1]:

$$\mathbf{E}(\mathbf{r}) = -j\omega\mu\mathbf{A}(\mathbf{r}) + \frac{1}{j\omega\epsilon} \nabla [\nabla \cdot \mathbf{A}(\mathbf{r})] \qquad (1.1)$$

where

$$\mathbf{A}(\mathbf{r}) = \iint_{S'} \mathbf{J}(\mathbf{r'})G(\mathbf{r}|\mathbf{r'}) \, \mathrm{d}s' \qquad (1.2)$$

and

$$G(\mathbf{r}|\mathbf{r'}) = \frac{e^{-jk|\mathbf{r}-\mathbf{r'}|}}{4\pi|\mathbf{r}-\mathbf{r'}|} \qquad (1.3)$$

In the equations above, $\mathbf{J}(\mathbf{r'})$ is the source current, $\mathbf{r'}$ is the source position vector, $\mathbf{A}(\mathbf{r})$ is the vector magnetic potential, and $\mathbf{r}$ is the vector to the observation point. The function $G(\mathbf{r}|\mathbf{r'})$ is the scalar Green's function which satisfies the equation

$$(\nabla^2 + k^2)G(\mathbf{r}|\mathbf{r'}) = -\delta(\mathbf{r} - \mathbf{r'}) \qquad (1.4)$$

In practice, the electric field is rarely expressed in the form shown above. An equivalent shorthand notation is generally used which allows the electric field to be expressed directly in terms of the source currents as

$$\mathbf{E(r)} = -jk\eta\left(\mathbf{I} + \frac{1}{k^2}\nabla\nabla\right) \cdot \iint_{S'} \mathbf{J(r')}\, G(\mathbf{r|r'})\, ds' \tag{1.5}$$

where

k = free-space wavenumber
η = free-space impedance

The odd-looking notation in (1.5) is easily translated into more conventional terms by simply comparing (1.5) with (1.1) and (1.2). The quantity $\mathbf{I}$ is referred to as the unit dyad (the identity matrix for our purposes), and the term $\nabla\nabla$ is a dyad which, when followed by a dot, represents the gradient of the divergence of a vector function. The knowledge that the terms $\mathbf{I}$ and $\nabla\nabla$ are "dyadic" operators is unnecessary; it is only important to know that (1.5) is equivalent to (1.1) and (1.2) and that the equivalence can be inferred by comparing the two representations for the electric field.

At this point, the usual far-field parallel rays approximation is made, transforming the Greens' function as

$$\frac{e^{-jk|\mathbf{r}-\mathbf{r'}|}}{4\pi|\mathbf{r}-\mathbf{r'}|} \rightarrow \frac{e^{-jkr}}{4\pi r}\, e^{jk\mathbf{r'}\cdot\hat{R}} \tag{1.6}$$

as shown in Figure 1.1.

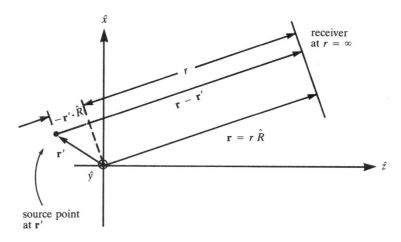

Figure 1.1 On the parallel rays approximation (source and receiver lie in the x-z plane).

The unit vector in the direction of the receiver is written in terms of the usual spherical coordinate angles (θ, ϕ) as

$$\hat{R} = \hat{x} \sin \theta \cos \phi + \hat{y} \sin \theta \sin \phi + \hat{z} \cos \theta \tag{1.7}$$

The radiation integral becomes

$$\mathbf{E}(\mathbf{r}) = -jk\eta \left(\mathbf{I} + \frac{1}{k^2} \nabla\nabla \right) \cdot \iint_{S'} \mathbf{J}(\mathbf{r}') \frac{e^{-jkr}}{4\pi r} e^{jkr' \cdot \hat{R}} \, ds' \tag{1.8}$$

and now the longitudinal (r) dependence of the Green's function has been separated from the transverse (θ, ϕ) dependence. Taking the observation region divergence operator underneath the integral sign, we see

$$\nabla \cdot \left[\mathbf{J}(\mathbf{r}') \frac{e^{-jkr}}{4\pi r} e^{jkr' \cdot \hat{R}} \right] = \mathbf{J}(\mathbf{r}') \cdot \nabla \left[\frac{e^{-jkr}}{4\pi r} e^{jkr' \cdot \hat{R}} \right] \tag{1.9}$$

In obtaining (1.9), we used the vector identity

$$\nabla \cdot (\psi \mathbf{a}) = \mathbf{a} \cdot \nabla\psi \tag{1.10}$$

valid when the vector $\mathbf{a}$ is constant (the source currents $\mathbf{J}(\mathbf{r}')$ are constant with respect to the observation coordinates). In spherical coordinates, the gradient operator is

$$\nabla = \hat{r} \frac{\partial}{\partial r} + \hat{\theta} \frac{1}{r} \frac{\partial}{\partial \theta} + \hat{\phi} \frac{1}{r \sin \theta} \frac{\partial}{\partial \phi} \tag{1.11}$$

so the r-component of the gradient in (1.9) is

$$\frac{1}{4\pi} \left[\frac{-jkre^{-jkr} - e^{-jkr}}{r^2} \right] e^{jkr' \cdot \hat{R}}$$

Retaining terms of order $1/r$ yields

$$-jk \frac{e^{-jkr}}{4\pi r} e^{jkr' \cdot \hat{R}}$$

for the r-component of the gradient in (1.9).

By (1.11), it is evident that the θ, ϕ components of the gradient vary with respect to r as

e^{-jkr}/r^2

By the factorized form of the Green's function (i.e., $G(r, \theta, \phi) = G_r(r) \cdot G_{\theta,\phi}(\theta, \phi)$), the (θ, ϕ) derivatives do not affect the r-dependence of the $(\hat{\theta}, \hat{\phi})$ components of the gradient. Therefore, in the far field,

$$\nabla\left[\frac{e^{-jkr}}{4\pi r} e^{jk\mathbf{r}'\cdot\hat{R}}\right] \rightarrow (-jk)\hat{R}\frac{e^{-jkr}}{4\pi r} e^{jk\mathbf{r}'\cdot\hat{R}} \tag{1.12}$$

We now apply the second differential operator in (1.8) (the gradient operator) to the dot product:

$$\mathbf{J}(\mathbf{r}') \cdot (-jk)\hat{R}\frac{e^{-jkr}}{4\pi r} e^{jk\mathbf{r}'\cdot\hat{R}}$$

$$= [\mathbf{J}(\mathbf{r}') \cdot \hat{R}]\left[-jk\frac{e^{-jkr}}{4\pi r} e^{jk\mathbf{r}'\cdot\hat{R}}\right] \tag{1.13}$$

$$= f(r, \theta, \phi) \cdot g(r, \theta, \phi)$$

Now

$$\nabla(fg) = f\nabla g + g\nabla f \tag{1.14}$$

We can easily evaluate the first term in (1.14) by (1.12):

$$f\nabla g = [\mathbf{J}(\mathbf{r}') \cdot \hat{R}] (-jk)^2\hat{R}\frac{e^{-jkr}}{4\pi r} e^{jk\mathbf{r}'\cdot\hat{R}} \tag{1.15}$$

By (1.7), the function $f = \mathbf{J}(\mathbf{r}') \cdot \hat{R} = J_r$ is only a function of (θ, ϕ). Therefore, by (1.11), only the $(\hat{\theta}, \hat{\phi})$ components of ∇f exist. As before, however, the $1/r$ factors contained in the $(\hat{\theta}, \hat{\phi})$ components of the gradient indicate that $g\nabla f$ varies as $1/r^2$. Thus, in the far field,

$$\nabla\left[\mathbf{J}(\mathbf{r}') \cdot (-jk)\hat{R}\frac{e^{-jkr}}{4\pi r} e^{jk\mathbf{r}'\cdot\hat{R}}\right] \rightarrow \hat{R}[\mathbf{J}(\mathbf{r}') \cdot \hat{R}] (-k^2)\frac{e^{-jkr}}{4\pi r} e^{jk\mathbf{r}'\cdot\hat{R}} \tag{1.16}$$

The far-field form of (1.8) now can be written as

$$\mathbf{E}^{rad}(\theta, \phi) = -jk\eta\frac{e^{-jkr}}{4\pi r} (\mathbf{I} - \hat{R}\hat{R}) \cdot \iint_{S'} \mathbf{J}(\mathbf{r}') e^{jk\mathbf{r}'\cdot\hat{R}} ds' \tag{1.17}$$

where the shorthand notation, $\hat{R}\hat{R} \cdot \mathbf{a}$, means that $\hat{R}(\hat{R} \cdot \mathbf{a})$. Equation (1.17) is the central equation of this book and the integral contained in it is the quantity we seek to evaluate by efficient numerical methods.

In evaluating the integral contained in (1.17), the currents are calculated using the *physical optics* approximation, which gives the current as

$$\mathbf{J}(\mathbf{r}') = 2\hat{n} \times \mathbf{H}^{\text{inc}}(\mathbf{r}') \tag{1.18}$$

where

$$\hat{n} = \text{unit outward normal from the reflector}$$
$$H^{\text{inc}} = \text{incident magnetic field on the reflector}$$

The current in (1.18) is the current that would be induced on an infinite conducting plane under illumination by an infinite plane wave. This approximation of the true currents ignores any reflections that might take place at the edges of the reflector. In addition, these currents do not satisfy the edge conditions, which require the normal component of current to be zero and the tangential component to be singular at an edge. It is remarkable that despite the apparent simplicity of this approximation, it yields results for the scattered field which are in close agreement with measured results, at least for the main beam region and the near side-lobes.

Directivity Calculations

We can see from (1.17) and (1.18) that the secondary scattered electric field strength is directly proportional to the incident field strength. If the incident field strength is already normalized with respect to some directivity standard (the usual standard being an isotropic radiator), the scattered far-field strength given by (1.17) will be normalized automatically with respect to the standard. If the incident field is not already normalized, the scattered far field must be correctly normalized.

For example, consider Figure 1.2 which illustrates schematically the problem of scattering by a reflector. The far zone E, H fields of the feed are related as

$$\mathbf{H}^{\text{inc}} = \frac{1}{\eta}\hat{\rho} \times \mathbf{E}^{\text{inc}} \tag{1.19}$$

hence

$$\mathbf{J}(\mathbf{r}') = \frac{2}{\eta}\hat{n} \times \hat{\rho} \times \mathbf{E}^{\text{inc}} \tag{1.20}$$

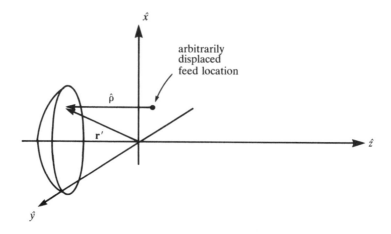

Figure 1.2 Reflector antenna geometry.

and

$$\mathbf{E}^{\text{rad}}(\theta, \phi) = -j/\lambda \, \frac{e^{-jkr}}{r} (\mathbf{I} - \hat{R}\hat{R}) \cdot \iint_{S'} (\hat{n} \times \hat{\rho} \times \mathbf{E}^{\text{inc}}) \, e^{jkr' \cdot \hat{R}} \, ds' \quad (1.21)$$

If $\mathbf{E}^{inc}$, the far-zone pattern of the feed, is not normalized, we calculate the directivity of the secondary scattered field as

$$D = \frac{|\mathbf{E}^{\text{rad}}(\theta, \phi)|^2}{\dfrac{1}{4\pi} \iint |\mathbf{E}^{\text{rad}}(\theta, \phi)|^2 \sin \theta \, d\theta \, d\phi} \quad (1.22)$$

where the integral in the denominator is taken over the complete sphere.

By conservation of energy, the denominator in (1.22) is equal to

$$\frac{1}{4\pi r^2} \cdot \{\text{the total power radiated by the feed}\}$$

Therefore,

$$D = \frac{4\pi r^2 |\mathbf{E}^{\text{rad}}(\theta, \phi)|^2}{\text{total feed power}} \quad (1.23)$$

This is the correct pattern directivity for the antenna.

Assuming that there is negligible reactive energy in the aperture field distribution, the maximum possible directivity obtainable from a given aperture can be easily calculated. The maximum directivity occurs when the aperture is uniformly illuminated; for example, if

$$|\hat{n} \times \hat{\rho} \times \mathbf{E}^{\text{inc}}| \, e^{jk\mathbf{r}' \cdot \hat{R}} \, ds' = r' \, dr' \, d\phi'$$

in (1.21), where r', ϕ' are polar coordinates in the circular projected aperture. Also, assume all of the energy radiated by the feed is intercepted by the reflector to produce the uniformly illuminated aperture. Then, the total electromagnetic power passing out of the circular exit aperture of the dish is equal to the area of the dish, or πa^2 (where a = radius of the dish). The uniformly illuminated aperture will have its main beam peak at $\theta = 0$ (boresight), so

$$D_{\text{max}} = \frac{4\pi r^2 \, |\mathbf{E}^{\text{rad}}(\theta = 0)|^2}{\text{total feed power}} \tag{1.24a}$$

or

$$D_{\text{max}} = \frac{(4\pi r^2) \, |\mathbf{E}^{\text{rad}}(\theta = 0)|^2}{\pi a^2} \tag{1.24b}$$

By (1.21)

$$|\mathbf{E}^{\text{rad}}(\theta = 0)| = \frac{1}{\lambda r} \cdot (\pi a^2) \tag{1.25}$$

Therefore,

$$D_{\text{max}} = \frac{(4\pi r^2)\left(\dfrac{\pi a^2}{\lambda r}\right)^2}{\pi a^2} \\ = \left(\frac{2\pi a}{\lambda}\right)^2 \tag{1.26}$$

This is the maximum possible directivity obtainable from a reflector antenna having a circular projected aperture of radius a.

Equation (1.23) gives the antenna directivity (commonly called *gain*) in terms of the total radiated feed power. One analytic feed pattern which is commonly used to approximate the pattern of a real feed is the $\cos^n$ feed pattern, where n is some real number. Assuming that this feed radiates only into the forward hemisphere, the total power radiated by the feed is given as

$$P_{\text{tot}} = \int_0^{\pi/2} \int_0^{2\pi} |\mathbf{E}(\theta'\phi')|^2 r^2 \sin\theta' \, d\theta' \, d\phi' \tag{1.27}$$

where,

$$r^2|E_\theta(\theta',\phi')| = \cos^m \theta' \cos\phi' \tag{1.28a}$$
$$r^2|E_\phi(\theta',\phi')| = \cos^n \theta' \sin\phi' \tag{1.28b}$$

for an $\hat{x}$-polarized feed. Thus,

$$r^2\{|E_\theta|^2 + |E_\phi|^2\} = \cos^{2m}\theta \cos^2\phi + \cos^{2n}\theta \sin^2\phi \tag{1.29}$$

Inserting (1.29) into (1.27) yields

$$P_{\text{tot}} = \pi \left[\frac{1}{2m+1} + \frac{1}{2n+1} \right] \tag{1.30}$$

as the total power radiated by the feed.

REFERENCES

1.   Harrington, R.F., *Time-Harmonic Electromagnetic Fields*, McGraw-Hill, New York, 1961.

Chapter 2
The Ludwig Method

One of the earliest techniques for evaluating efficiently the radiation integral in (1.17) was developed by Ludwig [1]. It is simple to understand and quite easy to implement as a computer program, although it is not nearly as efficient or mathematically sophisticated as some of the other methods of this book. Also, the generality of the method makes it a very versatile tool for analyzing a wide variety of reflector antennas. Lastly, the method has historical value as the predecessor of another well known integration technique—the quadratic phase method of Pogorzelski.

In the Ludwig approach, the three Cartesian components of the scattered field are calculated individually, reducing the vector integral to three scalar integrals. These three integrals must be evaluated at each angle (θ, ϕ) in space where the scattered electric field is to be determined. Therefore, the computer time required to generate an antenna pattern is directly proportional to the number of points used in generating the pattern.

To begin, cast the radiation integral in (1.17) into the more general form

$$E_u(\theta, \phi) = \iint_{S'} F(x', y', z' | \theta, \phi) \, ds', \, (u = x, y, z) \tag{2.1}$$

where $F(x', y', z' | \theta, \phi)$ is some complex valued function of the source coordinates x', y', z' and the far-field angles (θ, ϕ). If the function F is broken up into its magnitude and phase parts, and the projected aperture of the reflector in the x, y plane is assumed circular (not a necessary assumption), then

$$E_u(\theta, \phi) = \int_0^R \int_0^{2\pi} f(r', \phi') \, e^{j\psi(r', \phi')} \, dr' \, d\phi' \tag{2.2}$$

where R is the radius of the reflector in the x, y plane. The z' dependence is dropped since z' is assumed to be a known function of r', ϕ'.

Over some small sector of the reflector, we can approximate locally the magnitude and phase of the integrand as

$$f_i(r', \phi') = A_i + B_i(r' - r_i) + C_i(\phi' - \phi_i) \tag{2.3a}$$

$$\psi_i(r', \phi') = \alpha_i + \beta_i(r' - r_i) + \gamma_i(\phi' - \phi_i) \tag{2.3b}$$

where the coefficients are in general functions of θ, ϕ. Thus, (2.2) becomes

$$E_u(\theta, \phi) = \sum_i \iint [A_i + B_i(r' - r_i) + C_i(\phi' - \phi_i)]$$

$$\cdot e^{j[\alpha_i + \beta_i(r' - r_i) + \gamma_i(\phi' - \phi_i)]} \, dr' \, d\phi' \tag{2.4}$$

and finally,

$$E_u(\theta, \phi) = \sum_i \left\{ A_i e^{j\alpha_i} \int_{r_i}^{r_i + \Delta r_i} e^{j\beta_i(r' - r_i)} \, dr' \int_{\phi_i}^{\phi_i + \Delta \phi_i} e^{j\gamma_i(\phi' - \phi_i)} \, d\phi' \right.$$

$$+ B_i e^{j\alpha_i} \int_{r_i}^{r_i + \Delta r_i} (r' - r_i) e^{j\beta_i(r' - r_i)} \, dr' \int_{\phi_i}^{\phi_i + \Delta \phi_i} e^{j\gamma_i(\phi' - \phi_i)} \, d\phi' \tag{2.5}$$

$$\left. + C_i e^{j\alpha_i} \int_{r_i}^{r_i + \Delta r_i} e^{j\beta_i(r' - r_i)} \, dr' \int_{\phi_i}^{\phi_i + \Delta \phi_i} (\phi' - \phi_i) e^{j\gamma_i(\phi' - \phi_i)} \, d\phi' \right\}$$

All the integrals in (2.5) are easily evaluated. In fact, there are only two generic types:

Type 1

$$I_1 = \int_{u_1}^{u_1 + \Delta u} e^{j\xi(u - u_1)} \, du \tag{2.6}$$

and

Type 2

$$I_2 = \int_{u_1}^{u_1 + \Delta u} (u - u_1) e^{j\xi(u - u_1)} \, du \tag{2.7}$$

Using elementary techniques, we find

$$I_1 = \Delta u e^{j\xi\Delta u/2} \, \text{sinc} \, (\xi\Delta u/2) \tag{2.8}$$

where

$$\text{sinc} \, x = \sin x/x \tag{2.9}$$

and,

$$I_2 = j\frac{\Delta u^2}{2} e^{j3/4\xi\Delta u} \left\{ e^{-j\xi\Delta u/4} \left[\frac{\text{sinc} \, (\xi\Delta u/2) - 1}{(\xi\Delta u/2)} \right] - j \, \text{sinc} \, (\xi\Delta u/4) \right\} \tag{2.10}$$

where,

$$\frac{\text{sinc} \, x - 1}{x} \rightarrow -\frac{x}{6} \quad \text{as } x \rightarrow 0$$

The coefficients in (2.3) are chosen on a least squares basis, since each sector (other than the central pie-shaped sectors) has four corners, even though there are only three unknown coefficients (A, B, C or α, β, γ) to be determined for the sector. For example, A, B, C are chosen in such a way that the sum

$$[A_{ij} - f(r_i, \phi_j)]^2$$

$$+ [A_{ij} + B_{ij}(\Delta r_i) - f(r_i + \Delta r_i, \phi_j)]^2$$

$$+ [A_{ij} + C_{ij}(\Delta\phi_j) - f(r_i, \phi_j + \Delta\phi_j)]^2 \tag{2.11}$$

$$+ [A_{ij} + B_{ij}(\Delta r_i) + C_{ij}(\Delta\phi_j) - f(r_i + \Delta r_i, \phi_j + \Delta\phi_j)]^2 = \text{minimum}$$

where the i, j double subscript notation denotes a sector lying between r_i, $r_i + \Delta r_i$ and ϕ_j, $\phi_j + \Delta\phi_j$.

Equation (2.11) can be rewritten as

$$g(A_{ij}, B_{ij}, C_{ij}) = \text{minimum} \tag{2.12}$$

The minimum point occurs when

$$\frac{\partial g}{\partial A_{ij}} = \frac{\partial g}{\partial B_{ij}} = \frac{\partial g}{\partial C_{ij}} = 0 \tag{2.13}$$

Evaluating these three partial derivatives results in the following matrix equation

$$
\begin{bmatrix} 4 & 2 & 2 \\ 2 & 2 & 1 \\ 2 & 1 & 2 \end{bmatrix} \begin{bmatrix} A_{ij} \\ B_{ij}\,\Delta r_i \\ C_{ij}\,\Delta\phi_j \end{bmatrix} = \begin{bmatrix} f_{ij} + f_{i+1,j} + f_{i,j+1} + f_{i+1,j+1} \\ f_{i+1,j} + f_{i+1,j+1} \\ f_{i,j+1} + f_{i+1,j+1} \end{bmatrix} \tag{2.14}
$$

Solving this equation yields

$$
A_{ij} = \frac{3}{4} f_{ij} + \frac{1}{4} f_{i,j+1} + \frac{1}{4} f_{i+1,j} - \frac{1}{4} f_{i+1,j+1} \tag{2.15}
$$

$$
B_{ij}\Delta r_i = -\frac{1}{2} f_{ij} - \frac{1}{2} f_{i,j+1} + \frac{1}{2} f_{i+1,j} + \frac{1}{2} f_{i+1,j+1} \tag{2.16}
$$

$$
C_{ij}\,\Delta\phi_j = -\frac{1}{2} f_{ij} + \frac{1}{2} f_{i,j+1} - \frac{1}{2} f_{i+1,j} + \frac{1}{2} f_{i+1,j+1} \tag{2.17}
$$

For a parabolic reflector fed by a focused feed, and for observation angles within the main beam region, very few sample points will be required for good results. This is true, regardless of the size of the reflector in terms of wavelengths. Patches spanning several wavelengths may be quite adequate for pattern calculations. As the feed is defocused laterally (and to a lesser degree, axially), or if the pattern is calculated in the sidelobe region, the sample point density will increase, to a limit of perhaps 1 to 3 per wavelength on a side of a patch.

Ludwig [2] has presented some numerical results obtained using this algorithm. Figure 2.1 compares radiation patterns obtained both analytically and numerically (using the Ludwig algorithm) for the case of a 14λ diameter aperture with *parabolic squared* illumination. The plot indicates excellent agreement for 10 sidelobes.

As mentioned in [2], two major advantages of the Ludwig algorithm are that it degrades gracefully with increasing angle from the main beam, and that it can be adapted to an arbitrary grid over the reflector aperture. One final advantage for anyone engaged in developing advanced reflector antenna software is that this relatively simple Ludwig algorithm provides an excellent basis for debugging the more sophisticated antenna software.

Several papers [3–7] have appeared recently which compare the Ludwig algorithm with a number of other strictly numerical approaches to antenna pattern calculations. However, these purely numerical point-type integration schemes suffer from the appearance of non-physical grating lobes at wide angles [2]. Since the Ludwig algorithm is a patch-type algorithm, rather than a point-type, it is not susceptible to this problem.

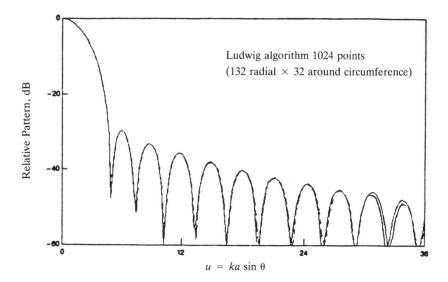**Figure 2.1** Radiation pattern of 14 λ diameter aperture (equivalent to Stutzman *et al.* [4, fig. 1]). (After [2] © 1988 IEEE.)

References

1. Ludwig, A.C., "Computation of Radiation Patterns Involving Numerical Double Integration," *IEEE Trans. Antennas Propagat.*, Vol. AP-16, pp. 767–769, November 1968.
2. Ludwig, A.C., "Comments on the Accuracy of the 'Ludwig' Integration Algorithm," *IEEE Trans. Antennas Propagat.*, Vol. AP-36, pp. 578–579, April 1988.
3. Lessow, H.A., W.V.T. Rusch, and H. Schjaer-Jacobsen, "On Numerical Evaluation of Two-Dimensional Phase Integrals," *IEEE Trans. Antennas Propagat.*, Vol. AP-23, pp. 714–717, September 1975.
4. Stutzman, W., S. Gilmore, and S. Stewart, "Numerical Evaluation of Radiation Integrals for Reflector Antenna Analysis Including a New Measure of Accuracy," *IEEE Trans. Antennas Propagat.*, Vol. AP-36, pp. 1018–1023, July 1988.
5. Craig, A.D., and P.D. Simms, "Fast Integration Techniques for Reflector Analysis," *Electron. Lett.*, Vol. 18, pp. 60–62, January 21, 1982.
6. Chugh, R.K., and L. Shafai, "Comparison of Romberg and Gauss Methods for Numerical Evaluation of Two-Dimensional Phase Integrals," *IEEE Trans. Antennas Propagat.*, Vol. AP-25, pp. 581–583, July 1977.
7. Lam, P.T., S.W. Lee, C.C. Hung, and R. Acousta, "Strategy for Reflector Pattern Calculation: Let the Computer Do the Work," *IEEE Trans. Antennas Propagat.*, Vol. AP-34, pp. 592–594, April 1986.

Chapter 3
Rusch's Method

The Ludwig method presented in Chapter 2 is a purely numerical approach to evaluating the radiation integral. One of the first approaches to an analytical evaluation of the radiation integral was presented by Rusch [1]. In this approach, the feed pattern and the reflector surface are expressed analytically, and the resulting radiation integral is also evaluated analytically. Historically, this method represents the beginning of modern reflector antenna analysis.

This method, while actually applicable to an arbitrary symmetrical reflector (with a laterally defocused feed), will be demonstrated here for the special case of a parabolic reflector. This will serve to keep the mathematics simple and to prevent the main ideas from being obscured by excessive analytical complexity.

To begin, consider the parabolic reflector shown in Figure 3.1 with on-axis feed. The reflector surface will be parameterized by the spherical coordinate angles (θ', ϕ') taken with respect to the origin at the focal point. The field points will be designated by the unprimed spherical coordinates (θ, ϕ). The radiated far field of the antenna can be written (by (1.17)) as

$$\mathbf{E}(\theta, \phi) = -jk\eta \, \frac{e^{-jkR}}{4\pi R} \, (\mathbf{I} - \hat{R}\hat{R}) \cdot \iint_{S'} \mathbf{J}(\mathbf{r}) \, e^{jk\rho' \cdot \hat{R}} \, \mathrm{d}s' \tag{3.1}$$

The elemental surface area on the reflector can be written (see Appendix A) as

$$\mathrm{d}s' = \frac{\sin \theta'' \, \mathrm{d}\theta'' \, \mathrm{d}\phi''}{\cos \dfrac{\theta''}{2} \, (1 + \cos \theta'')^2} \tag{3.2}$$

where $\theta'' = \pi - \theta'$ and $\phi'' = \phi'$.

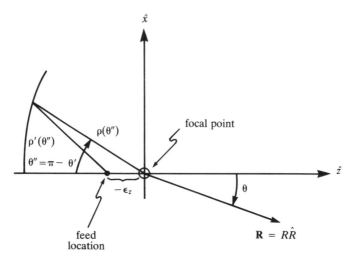

Figure 3.1 Schematic of a parabolic reflector with on-axis feed.

We can now parameterize the reflector in terms of θ'', ϕ'':

$$\rho(\theta'') = \frac{2F}{1 + \cos \theta''} \tag{3.3}$$

By the parallel rays approximation,

$$\rho'(\theta'') \cong \rho(\theta'') - \epsilon_z(\hat{z} \cdot \hat{\rho}) \tag{3.4}$$

For example,

$$\rho'(\theta'') \cong \frac{2F}{1 + \cos \theta''} + \epsilon_z \cos \theta'' \tag{3.5}$$

The phase term then becomes

$$\boldsymbol{\rho}'(\theta'') \cdot \hat{R} = \left[\frac{2F}{1 + \cos \theta''} + \epsilon_z \cos \theta'' \right] \hat{\rho} \cdot \hat{R} \tag{3.6}$$

where

$$\hat{\rho} \cdot \hat{R} = \sin \theta \sin \theta'' \cos (\phi'' - \phi) - \cos \theta \cos \theta'' \tag{3.7}$$

The exponential term thus takes the form

$$e^{jk\left[\frac{2F}{1\,+\,\cos\,\theta''}\,+\,\epsilon_z\,\cos\,\theta''\right]\left[\sin\,\theta\,\sin\,\theta''\,\cos\,(\phi''-\phi)\,-\,\cos\,\theta\,\cos\,\theta''\right]} \tag{3.8}$$

The current on the reflector is given as

$$\mathbf{J} = 2\hat{n} \times \mathbf{H}_f \tag{3.9}$$

where $\mathbf{H}_f$ is the magnetic field radiated by the feed. This can be expressed in terms of the radiated electric field of the feed as

$$\mathbf{H}_f = \frac{1}{\eta}\,\hat{\rho} \times \mathbf{E}_f \tag{3.10}$$

where η = free space impedance. So,

$$\mathbf{J} = \frac{2}{\eta}\,\hat{n} \times (\hat{\rho} \times \mathbf{E}_f) \tag{3.11}$$

Using the vector identity

$$\mathbf{a} \times (\mathbf{b} \times \mathbf{c}) = (\mathbf{a} \cdot \mathbf{c})\mathbf{b} - (\mathbf{a} \cdot \mathbf{b})\mathbf{c} \tag{3.12}$$

we get

$$\mathbf{J} = \frac{2}{\eta}\,[(\hat{n} \cdot \mathbf{E}_f)\hat{\rho} - (\hat{n} \cdot \hat{\rho})\mathbf{E}_f] \tag{3.13}$$

Referring to Figure A.2, we get

$$\hat{n} = -\cos\frac{\theta''}{2}\hat{\rho} + \sin\frac{\theta''}{2}\,\hat{\theta} \tag{3.14}$$

if we define

$$\hat{\theta} = \cos\theta''\,\cos\phi''\,\hat{x} + \cos\theta''\,\sin\phi''\hat{y} + \sin\theta''\hat{z} \tag{3.15}$$

Thus,

$$\hat{n} \cdot \hat{\rho} = -\cos\frac{\theta''}{2} \tag{3.16}$$

and

$$\hat{n} \cdot \mathbf{E}_f = E_{f,\theta} \sin \frac{\theta''}{2} \tag{3.17}$$

Hence,

$$\mathbf{J} = \frac{2}{\eta}\left[E_{f,\theta} \sin \frac{\theta''}{2}\hat{\rho} + \cos \frac{\theta''}{2}\mathbf{E}_f \right] \tag{3.18}$$

In rectangular coordinates this becomes

$$J_x = \frac{2}{\eta}\left[E_{f,\theta} \sin \frac{\theta''}{2} \sin \theta'' \cos \phi'' + \cos \frac{\theta''}{2}E_{fx} \right] \tag{3.19a}$$

$$J_y = \frac{2}{\eta}\left[E_{f,\theta} \sin \frac{\theta''}{2} \sin \theta'' \sin \phi'' + \cos \frac{\theta''}{2}E_{fy} \right] \tag{3.19b}$$

$$J_z = \frac{2}{\eta}\left[-E_{f,\theta} \sin \frac{\theta''}{2} \cos \theta'' + \cos \frac{\theta''}{2}E_{fz} \right] \tag{3.19c}$$

where

$$E_{fx} = E_{f,\theta} \cos \theta'' \cos \phi'' - E_{f,\phi} \sin \phi'' \tag{3.20a}$$

$$E_{fy} = E_{f,\theta} \cos \theta'' \sin \phi'' + E_{f,\phi} \cos \phi'' \tag{3.20b}$$

$$E_{fz} = E_{f,\theta} \sin \theta'' \tag{3.20c}$$

Substituting (3.20) into (3.19) yields

$$J_x = \frac{2}{\eta}\left[E_{f,\theta} \cos \frac{\theta''}{2} \cos \phi'' - \cos \frac{\theta''}{2} \sin \phi''E_{f,\phi} \right] \tag{3.21a}$$

$$J_y = \frac{2}{\eta}\left[E_{f,\theta} \cos \frac{\theta''}{2} \sin \phi'' + \cos \frac{\theta''}{2} \cos \phi''E_{f,\phi} \right] \tag{3.21b}$$

$$J_z = \frac{2}{\eta}\left[-E_{f,\theta} \sin \frac{\theta''}{2} \right] \tag{3.21c}$$

Now assume the feed to be linearly polarized in the $\hat{x}$-direction. For example,

$$E_{f,\theta} = \frac{e^{-jk\rho'}}{\rho'} a(\theta'') \cos \phi'' \qquad (3.22a)$$

$$E_{f,\phi} = -\frac{e^{-jk\rho'}}{\rho'} b(\theta'') \sin \phi'' \qquad (3.22b)$$

Note: a circularly polarized feed can be obtained by combining the $\hat{x}$-polarized feed pattern in (3.22) with a $\hat{y}$-polarized feed pattern that is in phase quadrature (i.e., multiplied by a factor of j). A $\hat{y}$-polarized feed pattern may be obtained from (3.22) by replacing the argument ϕ'' with $\phi'' - \pi/2$.

Substituting (3.22) into (3.21) gives

$$J_x = \frac{2}{\eta} \frac{e^{-jk\rho'}}{\rho'} \cos \frac{\theta''}{2} \left\{ \left[\frac{a(\theta'') + b(\theta'')}{2} \right] + \left[\frac{a(\theta'') - b(\theta'')}{2} \right] \cos 2\phi'' \right\} \qquad (3.23a)$$

$$J_y = \frac{2}{\eta} \frac{e^{-jk\rho'}}{\rho'} \cos \frac{\theta''}{2} \left[\frac{a(\theta'') - b(\theta'')}{2} \right] \sin 2\phi'' \qquad (3.23b)$$

$$J_z = -\frac{2}{\eta} \frac{e^{-jk\rho'}}{\rho'} a(\theta'') \sin \frac{\theta''}{2} \qquad (3.23c)$$

Now that the currents on the parabolic reflector have been expressed in terms of the coordinate variables (θ'', ϕ''), the ϕ'' integral in (3.1) can be evaluated in closed form. Using the fact that

$$\int_0^{2\pi} \left\{ \begin{array}{c} \cos m\phi \\ \sin m\phi \end{array} \right\} e^{jz\cos(\phi-\alpha)} \, d\phi = 2\pi j^m \left\{ \begin{array}{c} \cos m\alpha \\ \sin m\alpha \end{array} \right\} J_m(z) \qquad (3.24)$$

the radiation integrals

$$E_x = -j2k \frac{e^{-jkR}}{R} \int_0^{\theta_{max}} \int_0^{2\pi} \frac{e^{-jk\rho'(\theta'')}}{\rho'(\theta'')}$$

$$\cdot \left\{ \left[\frac{a(\theta'') + b(\theta'')}{2} \right] + \left[\frac{a(\theta'') - b(\theta'')}{2} \right] \cos 2\phi'' \right\}$$

$$\cdot e^{jk\left[\frac{2F}{1+\cos\theta''} + \epsilon_z\cos\theta'' \right][\sin\theta \sin\theta'' \cos(\phi'' - \phi) - \cos\theta \cos\theta'']} \qquad (3.25a)$$

$$\cdot \frac{\sin \theta''}{(1 + \cos \theta'')^2} \, d\theta'' \, d\phi''$$

$$E_y = -j2k\frac{e^{-jkR}}{R}\int_0^{\theta_{max}}\int_0^{2\pi}\frac{e^{-jk\rho'(\theta'')}}{\rho'(\theta'')}\cdot\left[\frac{a(\theta'')-b(\theta'')}{2}\right]\sin 2\phi''$$

$$\cdot\, e^{jk\left[\frac{2F}{1+\cos\theta''}+\epsilon_z\cos\theta''\right][\sin\theta\sin\theta''\cos(\phi''-\phi)-\cos\theta\cos\theta'']}$$

$$\cdot\,\frac{\sin\theta''}{(1+\cos\theta'')^2}\,d\theta''\,d\phi'' \tag{3.25b}$$

$$E_z = j2k\frac{e^{-jkR}}{R}\int_0^{\theta_{max}}\int_0^{2\pi}\frac{e^{-jk\rho'(\theta'')}}{\rho'(\theta'')}\,a(\theta'')\tan\frac{\theta''}{2}$$

$$\cdot\, e^{jk\left[\frac{2F}{1+\cos\theta''}+\epsilon_z\cos\theta''\right][\sin\theta''\sin\theta''\cos(\phi''-\phi)-\cos\theta\cos\theta'']}$$

$$\cdot\,\frac{\sin\theta''}{(1+\cos\theta'')^2}\,d\theta''\,d\phi'' \tag{3.25c}$$

become

$$E_x = -j2k\frac{e^{-jkR}}{R}\int_0^{\theta_{max}}\frac{e^{-jk\rho'(\theta'')}}{\rho'(\theta'')}\,e^{-jk\left[\frac{2F}{1+\cos\theta''}+\epsilon_z\cos\theta''\right]\cos\theta\cos\theta''}$$

$$\cdot\left\{2\pi\left[\frac{a(\theta'')+b(\theta'')}{2}\right]J_0\left[k\left(\frac{2F}{1+\cos\theta''}+\epsilon_z\cos\theta''\right)\sin\theta\sin\theta''\right]\right.$$

$$-\,2\pi\left[\frac{a(\theta'')-b(\theta'')}{2}\right]\cos 2\phi\, J_2\left[k\left(\frac{2F}{1+\cos\theta''}+\epsilon_z\cos\theta''\right)\right.$$

$$\left.\left.\cdot\sin\theta\sin\theta''\right]\right\}\cdot\frac{\sin\theta''}{(1+\cos\theta'')^2}\,d\theta'' \tag{3.26a}$$

$$E_y = j2k\frac{e^{-jkR}}{R}\int_0^{\theta_{max}}\frac{e^{-jk\rho'(\theta'')}}{\rho'(\theta'')}\left[\frac{a(\theta'')-b(\theta'')}{2}\right]$$

$$\cdot\,2\pi\sin 2\phi\, e^{-jk\left[\frac{2F}{1+\cos\theta''}+\epsilon_z\cos\theta''\right]\cos\theta\cos\theta''}$$

$$\cdot\, J_2\left[k\left(\frac{2F}{1+\cos\theta''}+\epsilon_z\cos\theta''\right)\sin\theta\sin\theta''\right]\cdot\frac{\sin\theta''}{(1+\cos\theta'')^2}\,d\theta'' \tag{3.26b}$$

$$E_z = j2k \frac{e^{-jkR}}{R} \int_0^{\theta_{\max}} \frac{e^{-jk\rho''(\theta'')}}{\rho'(\theta'')} a(\theta'') \tan\frac{\theta''}{2} \cdot e^{-jk\left[\frac{2F}{1+\cos\theta''}+\epsilon_z\cos\theta''\right]\cos\theta\cos\theta''}$$

$$\cdot 2\pi J_0\left[k\left(\frac{2F}{1+\cos\theta''}+\epsilon_z\cos\theta''\right)\sin\theta\sin\theta''\right] \cdot \frac{\sin\theta''}{(1+\cos\theta'')^2}\, d\theta'' \qquad (3.26c)$$

We note in passing that if $a(\theta'') = b(\theta'')$ (i.e., the feed has equal E- and H-plane patterns), then the x-component of the far-zone scattered field is independent of ϕ and the $\hat{y}$-component is zero. So, in the case of a high gain antenna having very narrow beamwidth, this condition produces a nearly perfect linearly-polarized radiation pattern. As might be expected, this type of "balanced feed" is critical to high performance antenna designs.

Note: The $\hat{z}$-component of the far-field pattern, though small, is still non-zero. Therefore, when the radial component of the radiation integral is subtracted, a small but finite amount of cross-polarization will result.

If it happens that the feed is both laterally and axially defocused, then the feed-to-dish distance may be expressed as

$$\rho'(\theta'', \phi'') = \rho(\theta'') - \boldsymbol{\epsilon} \cdot \hat{\rho}(\theta'', \phi'') \qquad (3.27)$$

where

$$\hat{\rho}(\theta'', \phi'') = \sin\theta''\cos\phi''\,\hat{x} + \sin\theta''\sin\phi''\,\hat{y} - \cos\theta''\,\hat{z} \qquad (3.28)$$

so,

$$\rho'(\theta'', \phi'') = [\rho(\theta'') + \epsilon_z\cos\theta''] - \epsilon_x\sin\theta''\cos\phi'' - \epsilon_y\sin\theta''\sin\phi'' \qquad (3.29)$$

Combining this term with the far-field parallel rays term

$$e^{jk\left[\frac{2F}{1+\cos\theta''}+\epsilon_z\cos\theta''\right]\sin\theta\sin\theta''(\cos\phi\cos\phi''+\sin\phi\sin\phi'')}$$

gives

$$e^{jk\sin\theta''(\epsilon_x\cos\phi''+\epsilon_y\sin\phi'')}$$

$$\cdot e^{jk\left[\frac{2F}{1+\cos\theta''}+\epsilon_z\cos\theta''\right]\sin\theta\sin\theta''(\cos\phi\cos\phi''+\sin\phi\sin\phi'')} = e^{jk\eta\sin\theta''\cos(\phi''-\alpha)} \qquad (3.30)$$

where,

$$\eta \cos \alpha = \epsilon_x + \left(\frac{2F}{1 + \cos \theta''} + \epsilon_z \cos \theta'' \right) \sin \theta \cos \phi \qquad (3.31a)$$

$$\eta \sin \alpha = \epsilon_y + \left(\frac{2F}{1 + \cos \theta''} + \epsilon_z \cos \theta'' \right) \sin \theta \sin \phi \qquad (3.31b)$$

Now substitute (3.30) into the radiation integrals and integrate as before.

This method can be extended to account for more general feed types by replacing the feed patterns in (3.22) with more general Fourier series. The one-term Fourier series in (3.22) apply for feeds having elliptical beams. Some feeds, such as rectangular horns, do not have this type of radiation pattern and will require the full Fourier series expansion.

The analysis presented in this chapter involves breaking up the current into its rectangular components and breaking up the far-zone scattered field into its rectangular components. Subtraction of the radial component is assumed throughout even though it is not explicitly stated. This Cartesian decomposition is quite effective for focused reflectors, but far less so for defocused reflectors such as hyperbolic subreflectors in Cassegrain systems. On the other hand, physical optics itself is less valuable in that instance and the *geometrical theory of diffraction* (GTD) is probably a more effective analysis tool for that problem. Rusch [1] and Wong [2] have both generalized this approach to the case of an arbitrary (focused or unfocused) symmetrical reflector and the reader is referred to the literature for further details.

As mentioned before, this method is somewhat math-intensive and perhaps most valuable as an historical milestone in antenna analysis, in light of many of today's current techniques. On the other hand, it represents the first analytical solution to the problem of physical optics scattering by a body-of-revolution reflector. This method should be compared with some of the method of moment solutions for scattering by bodies of revolution [3, 4].

Equation (3.26) indicates that for on-axis feeds, the radiation pattern can be represented by a Fourier series having at most two terms. In the case of laterally-defocused feeds, the required number of circumferential Fourier modes increases. A possible rule of thumb for determining the number of circumferential modes to include can be deduced by noting that the Bessel function decays harmonically when the argument is greater than the order. Because the order of the Bessel function equals the order of the circumferential harmonic, we may simply truncate the circumferential Fourier series at the integer equal to the maximum Bessel function argument.

REFERENCES

1. Rusch, W.V.T. and P.D. Potter, *Analysis of Reflector Antennas,* New York: Academic Press, 1970.
2. W.C. Wong, "On the Equivalent Parabola Technique to Predict the Performance Characteristics of a Cassegrainian System with an Offset Feed," *IEEE Trans. Antennas Propagat.,* Vol. AP-21, pp. 335–339, May 1973.
3. M.G. Andreason, "Scattering from Bodies of Revolution," *IEEE Trans. Antennas Propagat.,* Vol. AP-13, pp. 303–310, March 1965.
4. Mautz, J.R. and R.F. Harrington, "Radiation and Scattering from Bodies of Revolution," *Appl. Sci. Res.,* Vol. 20, pp. 405–435, 1969.

Suggestions for Additional Reading

1. Imbriale, W.A., P.G. Ingerson, and W.C. Wong, "Large Lateral Feed Displacements in a Parabolic Reflector," *IEEE Trans. Antennas Propagat.,* Vol. AP-22, pp. 742–745, November 1974.

Chapter 4
The Jacobi-Bessel Method

The Jacobi-Bessel approach was first introduced in a landmark 1977 paper by Galindo-Israel and Mittra [1]. Their revolutionary approach to reflector antenna analysis is based in part on work performed much earlier by Zernike [2] in connection with the theory of aberrations in optical systems. Thus, their Jacobi-Bessel method actually integrates classical optics with modern antenna theory.

The Jacobi-Bessel method, though mathematically complex, is actually quite easy to understand conceptually. It first involves separating the aperture and far-field variables in the radiation integral. Next, the slowly varying aperture distribution is expressed as a sum over a set of orthogonal functions defined on the circular antenna aperture. Finally, each of the basis functions is integrated against the Fourier kernel over the circular aperture. These integrals are closed-form functions of the far-field coordinates, hence the radiation integral can be expressed directly in terms of the far-field angles and the aperture expansion coefficients. This scheme, along with a few extra details to account for reflector curvature, forms the basis of the Jacobi-Bessel method.

It is worth noting that there are two conflicting considerations which affect the choice of aperture basis functions in the scheme described above: the choice of a basis set that best represents the aperture distribution with the least number of functions and, the choice of a basis set that is easily integrable against the Fourier kernel. In this method, ease of integrability is perhaps the primary concern. In Chapter 5 we will consider the Fourier-Bessel method, which employs an entirely different set of aperture expansion functions.

To examine the Jacobi-Bessel method in greater detail, consider Figure 4.1. As always, the radiation integral is given as

$$\mathbf{E}(\mathbf{r}) = -jk\eta\frac{e^{-jkR}}{4\pi R}(\mathbf{I} - \hat{R}\hat{R}) \cdot \iint_{S'}\mathbf{J}(\mathbf{r}')e^{jk\hat{R}\cdot\mathbf{r}'}\,ds' \tag{4.1}$$

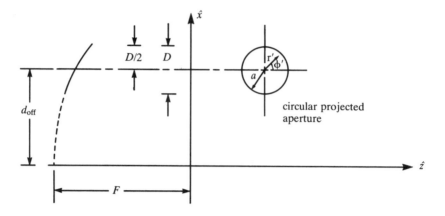

Figure 4.1 Offset parabolic reflector.

The integration is performed with respect to an origin located at the center of the circular projected aperture of the reflector. If we let

$$ds' = r'dr'd\phi' \cdot J_{\Sigma}$$

where (r', ϕ') are polar coordinates in the circular projected aperture and J_{Σ} is the surface Jacobian given by $|\hat{n} \cdot \hat{z}|^{-1}$, then the equation for the radiated electric field becomes

$$\mathbf{E}(\mathbf{r}) = -j\frac{\eta}{2\lambda}\frac{e^{-jkR}}{R}(\mathbf{I} - \hat{R}\hat{R}) \cdot \int_0^a\int_0^{2\pi}\mathbf{J}(\mathbf{r}')J_{\Sigma}e^{jk\hat{R}\cdot\mathbf{r}'}r'dr'd\phi' \qquad (4.2)$$

The vectors $\hat{R}$, $\mathbf{r}'$ in the exponent are given as

$$\begin{aligned}\hat{R} &= \sin\theta\cos\phi\hat{x} + \sin\theta\sin\phi\hat{y} + \cos\theta\hat{z}\\ &= u\hat{x} + v\hat{y} + w\hat{z}\end{aligned} \qquad (4.3)$$

and,

$$\mathbf{r}' = x'\hat{x} + y'\hat{y} + z'\hat{z} \qquad (4.4)$$

Note: u, v, w will prove far more convenient to work with than (θ, ϕ).

The source coordinates can be written in terms of r', ϕ' as

$$x' = r'\cos\phi' + d_{\text{off}} \qquad (4.5)$$

$$y' = r' \sin \phi' \tag{4.6}$$

$$z' = \frac{1}{4F}(x'^2 + y'^2) - F$$

$$= \frac{1}{4F}[r'^2 + (2\,d_{\text{off}})\,r' \cos \phi'] + \left[\frac{d_{\text{off}}^2 - 4F^2}{4F}\right] \tag{4.7}$$

where
d_{off} is the aperture offset in the x direction.

The dot product of $\hat{R}$, r' can now be written as

$$\hat{R} \cdot r' = r' \left\{ \left[(u - u_0) + \frac{d_{\text{off}}}{2F}(w - w_0)\right] \cos \phi' + (v - v_0) \sin \phi' \right\}$$

$$+ u \cdot d_{\text{off}} + \frac{w}{4F}(d_{\text{off}}^2 - 4F^2)$$

$$+ r'\left[\left(u_0 + \frac{d_{\text{off}} \cdot w_0}{2F}\right) \cos \phi' + v_0 \sin \phi'\right]$$

$$+ \frac{1}{4F}(r'^2 - a^2)(w - w_0) + \frac{1}{4F}(r'^2 - a^2)w_0 + \frac{1}{4F}a^2 w \tag{4.8}$$

where u_0, v_0, w_0 are the far-field coordinates of the expected main beam position.

The change of variables $r' = as$ (where s ranges between 0 and 1) gives

$$\mathbf{E}(\mathbf{r}) = -j\frac{\eta a^2}{2\lambda}\frac{e^{-jkR}}{R} e^{jk d_{\text{off}} u} e^{jk\left(\frac{d_{\text{off}}^2 - 4F^2}{4F}\right) - w} e^{jk(a^2/4F)w}$$

$$\cdot \int_0^1 \int_0^{2\pi} \mathbf{J}(s, \phi') J_\Sigma\, e^{jk(a^2/4F)(s^2 - 1)w_0}\, e^{jkas\left[\left(u_0 + \frac{d_{\text{off}} w_0}{2F}\right)\cos\phi' + v_0 \sin\phi'\right]}$$

$$\cdot e^{jk(a^2/4F)(s^2 - 1)(w - w_0)}\, e^{jkas\left\{\left[(u - u_0) + \frac{d_{\text{off}}}{2F}(w - w_0)\right]\cos\phi' + (v - v_0)\sin\phi'\right\}}\, s\, ds\, d\phi' \tag{4.9}$$

where subtraction of the radial component is assumed. The integral portion of (4.9) can now be rewritten in the following simplified form:

$$\mathbf{I} = \int_0^1 \int_0^{2\pi} \mathbf{f}(s, \phi')\, e^{jk(a^2/4F)(s^2 - 1)(w - w_0)}\, e^{jkas(\bar{u}\cos\phi' + \bar{v}\sin\phi')}\, s\, ds\, d\phi' \tag{4.10}$$

Some important features of (4.10) should be mentioned before proceeding further. The right hand side of this equation is a highly factored form of the original radiation integral. The first term, $\mathbf{f}(s, \phi')$, is the effective aperture distribution consisting of the surface current, $\mathbf{J}$, the surface Jacobian, as well as various other factors which depend solely on the source coordinates (s, ϕ'). The third term in the integrand is a Fourier kernel involving far-field coordinates $(\bar{u}, \bar{v})$, which in general are not equal to $u = \sin\theta\cos\phi$, $v = \sin\theta\sin\phi$. If the radiation integral were limited to these two terms, reflector antenna analysis would be easy.

Unfortunately, the radiation integral in (4.10) also contains an additional factor, which accounts for the finite depth of the reflector surface. This term has been factored into a very special form, which will later be expanded into a Taylor series in the complex variable

$$z = jk \frac{a^2}{4F} (s^2 - 1)(w - w_0)$$

The Taylor series will naturally converge best when z lies near the origin in the complex plane. Clearly, z will be small in the vicinity of the main lobe direction since the quantity $(w - w_0)$ will be small there also. Away from the main lobe region (i.e., away from any geometrical optics reflected rays), the antenna pattern will be governed primarily by fields diffracted at the edge of the reflector. Thus, currents near the edge of the reflector will constitute the primary contribution to the radiation integral. However, z will also be small near the reflector rim, due to the presence of the $(s^2 - 1)$ term. Therefore, this particular factorization of z ensures that z will be small over a wide range of far-field angles and that the exponential can be approximated by a Taylor series of minimum length.

Rewrite (4.10) in the form

$$I = \int_0^1 \int_0^{2\pi} \mathbf{f}(s, \phi') \, e^{jka^2/4F(s^2-1)(w-w_0)} \, e^{jkas\eta\cos(\phi'-\alpha)} s \, ds \, d\phi' \qquad (4.11)$$

where

$$\bar{u} = \eta\cos\alpha \qquad (4.12a)$$
$$\bar{v} = \eta\sin\alpha \qquad (4.12b)$$

The Taylor series expansion for the exponential is

$$e^z = \sum_{p=0}^P \frac{z^p}{p!} \quad (z \text{ complex}) \qquad (4.13)$$

So, the previous integral becomes

$$I_u = \sum_{p=0}^{P} \frac{1}{p!} \left(jk \frac{a^2}{4F} \right)^p (w - w_0)^p$$

$$\int_0^1 \int_0^{2\pi} (s^2 - 1)^p f_u(s, \phi') \, e^{jkas\eta\cos(\phi' - \alpha)} \, sdsd\phi' \tag{4.14}$$

where u represents x, y, z scalar components.

We now expand each component function f_u in terms of a set of orthogonal functions defined on the unit disk. A Fourier series will be used in the circumferential direction. At this point, we will not yet specify the radial functions other than to say that they will be some two-parameter set of functions (such as the associated Legendre functions, for example). Later it will become clearer as to the kinds of properties we want the radial functions to possess.

So, expand each component function f_u (s, ϕ') as follows

$$f_u(s, \phi') = \sum_{m=0}^{M} \sum_{n=0}^{N} [C_m^n \cos n\phi' + D_m^n \sin n\phi'] F_m^n(s) \tag{4.15}$$

where the radial functions $F_m^n(s)$ are yet to be specified. The integral now becomes

$$I_u = \sum_{p=0}^{P} \frac{1}{p!} \left(jk \frac{a^2}{4F} \right)^p (w - w_0)^p$$

$$\cdot \sum_{m=0}^{M} \sum_{n=0}^{N} C_m^n \, {}^p I_{m,\cos}^n + D_m^n \, {}^p I_{m,\sin}^n \tag{4.16}$$

where

$$^p I_{m\{\cos\}}^n {}^{\{\sin\}} = \int_0^1 \int_0^{2\pi} (s^2 - 1)^p F_m^n(s) \begin{Bmatrix} \sin n\phi' \\ \cos n\phi' \end{Bmatrix} e^{jkas\eta\cos(\phi' - \alpha)} s \, ds \, d\phi' \tag{4.17}$$

We will examine the $p = 0$ integrals first and then hopefully calculate the others by recursion, based on the properties of the F_m^n (s). The ϕ'-integrals can be easily evaluated as

$$^0 I_{m\{\sin\}}^n {}^{\{\cos\}} = 2\pi j^n \begin{Bmatrix} \cos n\alpha \\ \sin n\alpha \end{Bmatrix} \int_0^1 F_m^n(s) J_n(kas\eta) s \, ds \tag{4.18}$$

At this point, it is now necessary to specify the radial functions, in order to evaluate the radial portion of the integral. Clearly, one criterion must be to choose the F_m^n (s) so that the integral in (4.18) can be evaluated in closed form. Galindo-Israel and Mittra use a set of functions they term *modified Jacobi polynomials* which they relate to the Jacobi polynomials $P_m^{(n,0)}$ $(\cdot)$ according to the following equation:

$$F_m^n(s) = \sqrt{2(n + 2m + 1)}P_m^{(n,0)}(1 - 2s^2)s^n \qquad (4.19)$$

Galindo-Israel and Mittra also use the abbreviated notation P_m^n $(1 - 2s^2)$ to represent the right-hand side of (4.19).

An in-depth study of Jacobi polynomials would be outside the scope of this chapter, so their relevant mathematical properties only will be discussed here. For additional information, the reader is directed to the literature [3, 4].

The orthogonality property of the modified Jacobi polynomials is stated in [1] as

$$\int_0^1 F_m^n(s) \, F_{m'}^n(s) \, s \, ds = \delta_{mm'} \qquad (4.20)$$

where δ_{mn} is the Kronecker delta,

$$\begin{aligned} \delta_{mn} &= 1 \, (m = n) \\ &= 0 \, (m \neq n) \end{aligned} \qquad (4.21)$$

Equation (4.20) can be deduced from the orthogonality property of the Jacobi polynomials, $P_n^{(\alpha,\beta)}$, which is given in [3, equation 22.1.2]:

$$\int_{-1}^1 \left[P_n^{(\alpha,0)} (x) \right]^2 (1 - x)^\alpha \, dx = \frac{2^{\alpha+1}}{2n + \alpha + 1} \qquad (4.22)$$

where α is an integer. The change of variables $x = 1 - 2 s^2$ yields

$$dx = -4s \, ds \qquad (4.23)$$

and

$$(1 - x)^\alpha = (2s^2)^\alpha \qquad (4.24)$$

The integral (4.22) becomes

$$\int_1^0 [P_n^{(\alpha,0)}(1 - 2s^2)]^2 2^\alpha \, s^{2\alpha} \, (-4 \, s \, ds) = \frac{2^{\alpha+1}}{2n + \alpha + 1} \qquad (4.25)$$

Reversing the order of integration gives

$$\int_0^1 \left[\sqrt{2(\alpha + 2n + 1)}\, s^\alpha\, P_n^{(\alpha,0)}\, (1 - 2s^2) \right]^2 sds = 1 \tag{4.26}$$

which is the same as (4.20). Also,

$$\int_0^1 F_n^\alpha(s)\, F_m^\alpha(s)\, sds = 0 \tag{4.27}$$

by [3, equation 22.1.1].

The purpose of this brief exercise has been to de-mystify the modified Jacobi polynomials and to relate them to the ordinary Jacobi polynomials from orthogonal polynomial theory. The orthogonality properties allow the C_m^n, D_m^n coefficients from (4.15) to be solved for as

$$\begin{Bmatrix} C_m^n \\ D_m^n \end{Bmatrix} = \frac{\epsilon_n}{2\pi} \int_0^1 s\, ds \int_0^{2\pi} d\phi'\, f_u(s, \phi') \begin{Bmatrix} \cos n\phi' \\ \sin n\phi' \end{Bmatrix} F_m^n(s) \tag{4.28}$$

where ϵ_n is the Neumann number defined as

$$\begin{aligned} \epsilon_n &= 1 \quad (n = 0) \\ &= 2 \quad (n \neq 0) \end{aligned} \tag{4.29}$$

The modified Jacobi polynomials can also be integrated against the Bessel function in closed form [1] as follows:

$$\int_0^1 F_m^n(s)\, J_n(kas\eta)\, s\, ds = \sqrt{2(n + 2m + 1)}\, \frac{J_{n+2m+1}(ka\eta)}{(ka\eta)} \tag{4.30}$$

This allows the integral in (4.18) to be expressed as

$$^0I_m^n {\cos \atop \sin} = 2\pi j^n \begin{Bmatrix} \cos n\alpha \\ \sin n\alpha \end{Bmatrix} \sqrt{2(n + 2m + 1)}\, \frac{J_{n+2m+1}(ka\eta)}{(ka\eta)} \tag{4.31}$$

As mentioned earlier, our strategy is to evaluate the $p = 0$ integrals in (4.18) and (4.31) in closed form and then to evaluate the others by recursion. Before proceeding, it is first necessary to study the recursion properties of the modified Jacobi polynomials. We will begin by stating the relevant recursion relation for the modified Jacobi polynomials and then deriving it from the well known recursion relation for ordinary orthogonal Jacobi polynomials. The recursion relation we seek is

$$(s^2 - 1)\, F_m^n = a_{mn}\, F_{m-1}^n + b_{mn} F_m^n + c_{mn} F_{m+1}^n \tag{4.32}$$

where

$$a_{mn} = -\frac{d_m}{d_{m-1}}\left[\frac{m\,(m + n)}{(n + 2m)(n + 2m + 1)}\right] \tag{4.33}$$

$$b_{mn} = \frac{n^2 - (n + 2m)(n + 2m + 2)}{2\,(n + 2m)(n + 2m + 2)} \tag{4.34}$$

$$c_{mn} = -\frac{d_m}{d_{m+1}}\left[\frac{(m + 1)(n + m + 1)}{(n + 2m + 1)(n + 2m + 2)}\right] \tag{4.35}$$

$$d_m = \sqrt{2(n + 2m + 1)} \tag{4.36}$$

This recurrence can be derived easily from the well-known recurrence relation for Jacobi polynomials [3, 4]:

$$2(n + 2m)(m + 1)(n + m + 1)\, P_{m+1}^{(n,0)}(x)$$
$$= (n + 2m + 1)[(n + 2m)(n + 2m + 2)x + n^2]\, P_m^{(n,0)}(x)$$
$$- 2m\,(m + n)(n + 2m + 2)\, P_{m-1}^{(n,0)}(x) \tag{4.37}$$

Making the change of variable, $x = 1 - 2s^2$, and then expressing the Jacobi polynomial in terms of the modified Jacobi polynomial produces (after some math) the recurrence defined by (4.32–4.36). Multiplying (4.32) by $(s^2 - 1)^{p-1}$ gives

$$(s^2 - 1)^p F_m^n(s) = a_{mn}\, (s^2 - 1)^{p-1}\, F_{m-1}^n(s)$$
$$+ b_{mn}(s^2 - 1)^{p-1}\, F_m^n\,(s)$$
$$+ c_{mn}(s^2 - 1)^{p-1}\, F_{m+1}^n\,(s) \tag{4.38}$$

Integrating this equation as in (4.17) yields immediately

$$^p I_m^n \begin{Bmatrix} \cos \\ \sin \end{Bmatrix} = a_{mn} \,^{p-1} I_{m-1}^n \begin{Bmatrix} \cos \\ \sin \end{Bmatrix}$$
$$+ b_{mn} \,^{p-1} I_m^n \begin{Bmatrix} \cos \\ \sin \end{Bmatrix}$$
$$+ c_{mn} \,^{p-1} I_{m+1}^n \begin{Bmatrix} \cos \\ \sin \end{Bmatrix} \tag{4.39}$$

With this equation, we can calculate the integrals for all values of p, knowing only the integral for $p = 0$. The only problem that might arise in using this recursion would be when $m = 0$, since it is not obvious what form the Jacobi polynomials take when the order is negative (i.e., when $m - 1 < 0$). Fortunately, the integrals for all orders of p can be computed directly when $m = 0$ since

$$P_0^{(n,0)}(x) = 1 \qquad (4.40)$$

and

$$F_0^n(s) = \sqrt{2(n + 1)}\, s^n \qquad (4.41)$$

With a little effort, it can be shown that

$$\int_0^1 (s^2 - 1)^p\, s^{n+1}\, J_n(ka\eta s)\ ds$$

$$= (-2)^p\, p!\, \frac{J_{n+p+1}(ka\eta)}{(ka\eta)^{p+1}} \qquad (4.42)$$

so that

$$^pI^n{}_{0\{{\cos \atop \sin}\}} = 2\pi j^n \left\{ {\cos n\alpha \atop \sin n\alpha} \right\} \sqrt{2(n + 1)}$$

$$\cdot (-2)^p\, p!\, \frac{J_{n+p+1}(ka\eta)}{(ka\eta)^{p+1}} \qquad (4.43)$$

where $p! = p(p - 1)(p - 2) \ldots (2)(1)$.

The integral represented in (4.14) now becomes

$$I_u = \sum_{p=0}^{P} \frac{1}{p!}\left(jk\frac{a^2}{4F}\right)^p (w - w_0)^p$$

$$\cdot \sum_{m=0}^{M} \sum_{n=0}^{N} [C_m^n\, {}^pI_{m,\cos}^n + D_m^n\, {}^pI_{m,\sin}^n] \qquad (4.44)$$

and

$$E_u(r) = -j\frac{\eta a^2}{2\lambda}\frac{e^{-jkR}}{R}\, e^{jkd_{\text{off}}u}\, e^{jk\left(\frac{d_{\text{off}}^2 - 4F^2}{4F}\right)w}$$

$$\cdot \, e^{jk\frac{a^2}{4F}w} \sum_{p=0}^{P} \frac{1}{p!} \left(jk\frac{a^2}{4F}\right)^p (w - w_0)^p \left\{ \sum_{m=0}^{M} \sum_{n=0}^{N} C_m^n \, {}^P I_{m,\cos}^n \right. \qquad (4.45)$$

$$\left. + \, D_m^n \, {}^P I_{m,\sin}^n \right\}$$

This completes the analysis of the Jacobi-Bessel method. Note that the ${}^P I_m^n$ integrals in (4.45) are independent of polarization—only the C, D coefficients depend on the current and far-zone electric field vector components.

In the integrals (4.28) for the C_m^n, D_m^n coefficients, the Jacobi polynomials are easily calculated using the recursion (4.32) and the special values [3]

$$P_0^{(n,0)} = 1$$

$$P_1^{(n,0)} = \frac{1}{2}[n + (n + 2)x]$$

One convenient way to evaluate the double integrals is to use the *fast Fourier transform* (FFT) for the circumferential integrals and then use Gaussian quadrature to integrate the circumferential integrals in the radial direction. For reference, the first few modified Jacobi polynomials are shown in Figure 4.2 [1].

The Jacobi-Bessel method can also be adapted to reflectors having elliptical projected apertures [5]. The primary analytical differences are 1) that the actual spatial coordinates x', y' are expressed in terms of dimensionless coordinates (t, ψ) such that

$$x' = at \cos \psi \qquad (4.46a)$$

$$y' = bt \sin \psi \qquad (4.46b)$$

(a, b) = half-diameters of the ellipse

and 2) that the z-dependence of the dish depth term is not expressed in terms of the projected aperture coordinates. It is left in the form

$$e^{jkz'(w - w_0)}$$

Next, this term is factored as

$$e^{jkz_c(w - w_0)} \, e^{jk(z' - z_c)(w - w_0)}$$

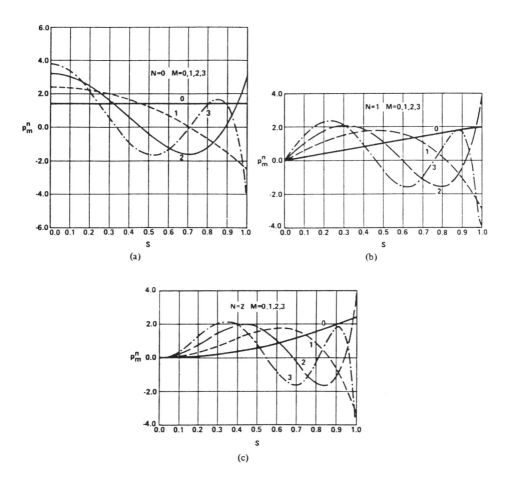

Figure 4.2 Modified Jacobi polynomials (after Galindo-Israel [1]).(© 1979 IEEE.)

and the left-hand term is factored out of the radiation integral. The planar surface $z' = z_c$ is somewhat arbitrary, but as mentioned in [5], it must be chosen as close as possible to the reflector rim for improved convergence behavior of the Taylor series expansion.

Figure 4.3 shows the convergence properties of the Jacobi-Bessel series for a 50λ circular reflector. In this case, only the $p = 0$ term of the Taylor series for the exponential is used. It is seen that convergence is quite good, even for rather wide angle scanning.

We note from the preceding discussion that the explicit form of the Jacobi polynomials still has not been given. This form is not at all complex and is given by either of the following two expressions [3]:

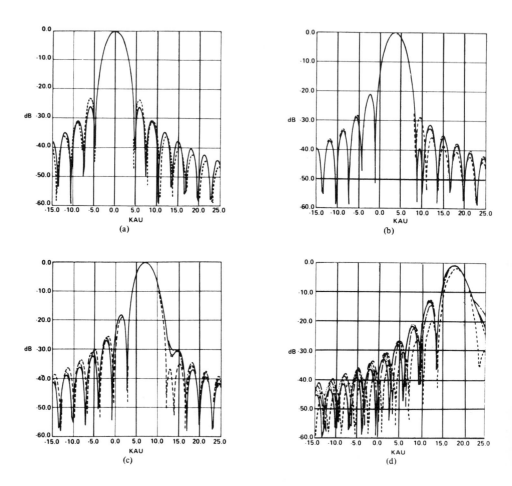

Figure 4.3 Series convergence as function of scan. $f/D = 0.5$, $D/\lambda = 50$, $kau_0 = kau_m$, $ka\eta_e = 5$, $\cos^n \theta'$ feed, -10dB taper. (a) $kau_m = 0$. (b) $kau_m = 3.5$, $kau_s = -5$. (c) $kau_m = 7$, $kau_s = -8$. (d) $kau_m = 17.5$, $kau_s = -20$. —— converged. ----- $M = N = 2$. = $M = N = 3$. $-$ = $=$ $-$ = $=$ $-$ $M = N = 4$. ..... $M = N = 5$. (After Galindo-Israel [1].) (© 1979 IEEE.)

$$P_n^{(\alpha,0)}(x) = \frac{1}{2^n} \sum_{m=0}^{n} \binom{n+\alpha}{m}\binom{n}{n-m}(x-1)^{n-m}(x+1)^m \tag{4.47}$$

or,

$$P_n^{(\alpha,0)}(x) = \frac{1}{n!} \sum_{m=0}^{n} \binom{n}{m}\frac{(\alpha+n+m)!}{2^m(\alpha+m)!}(x-1)^m \tag{4.48}$$

where α is an integer greater than or equal to zero and

$$\binom{n}{m} = \frac{n!}{m! \, (n - m)!} \qquad (4.49)$$

where

$$n! = n \, (n - 1)(n - 2) \, . \, . \, . \, (2)(1) \qquad (4.50)$$

The Gaussian integration scheme proposed for the radial integration is somewhat inefficient since it results in a large number of numerical integration points clustered near the center point of the circular projected aperture. This happens because in an ordinary Gaussian integration scheme the sample points for the integrand tend to cluster near the ends of the integration interval. One simple way to fix this problem is to perform the radial Gaussian integration across the entire diameter of the reflector, rather than from the center of the dish out to the radius. The circumferential integration limits would also be changed from $(0, 2\pi)$ to $(0, \pi)$. A much more elegant approach to this problem has been proposed by Jamnejad [6, 7]. In this approach, the same orthogonal functions which are used for the aperture expansion are used for the numerical integration of the aperture expansion coefficients.

REFERENCES

1.   Galindo-Israel, V. and R. Mittra, "A New Series Representation for the Radiation Integral with Application to Reflector Antennas," *IEEE Trans. Antennas Propagat.*, Vol. AP-25, pp. 631–635, September 1977. (Correction, *IEEE Trans. Antennas Propagat.*, Vol. AP-26, p. 628, July 1978.)
2.   Zernike, F., *Physica*, 1 (1934), p. 689.
3.   Abramowitz, M. and I.A. Stegun, eds., *Handbook of Mathematical Functions*. New York: Dover, 1970.
4.   Beckman, P., *Orthogonal Polynomials for Engineers and Physicists*. Boulder: Golem Press, 1973.
5.   Rahmat-Samii, Y., "Jacobi-Bessel Analysis of Reflector Antennas with Elliptical Apertures," *IEEE Trans. Antennas Propagat.*, Vol. AP-35, pp. 1070–1074, September 1987.
6.   Jamnejad, V., "A New Look at the Circle Polynomial Series Representation of the Radiation Integral of Reflector Antennas," *Antennas and Propagation International Symposium Digest*, Vol. 1, pp. 380–384, June 1984.
7.   Jamnejad, V., "A New Integration Scheme for Application to the Analysis of Reflector Antennas," Antennas and Propagation International Symposium Digest, Vol. 1, pp. 380–384, June 1984.

Suggestions for Additional Reading

1. Mittra, R., Y. Rahmat-Samii, V. Galindo-Israel, and R. Norman, "An Efficient Technique for the Computation of Vector Secondary Patterns of Offset Paraboloid Reflectors," *IEEE Trans. Antennas Propagat.*, Vol. AP-27, pp. 294–304, May 1979.
2. Galindo-Israel, V. and Y. Rahmat-Samii, "A New Look at Fresnel Field Computation Using the Jacobi-Bessel Series," *IEEE Trans. Antennas Propagat.*, Vol. AP-29, pp. 885–898, November 1981.
3. Rahmat-Samii, Y. and V. Galindo-Israel, "Shaped Reflector Analysis using the Jacobi-Bessel Series," *IEEE Trans. Antennas Propagat.*, Vol. AP-28, pp. 425–435, July 1980.

Chapter 5
The Fourier-Bessel Method

The previous Chapter considered a particular implementation of a general approach to reflector antenna analysis. To reiterate, this approach centers around expanding the aperture distribution as a sum over a set of orthogonal functions. Each of these functions is then integrated in closed form, allowing the radiation integral to be expressed explicitly in terms of the aperture expansion coefficients and the far-field variables. This scheme forms the basis of most modern reflector antenna analysis techniques.

This straightforward scheme can be implemented in a variety of ways. The implementation used in Chapter 4 involved some rather complex mathematics. The implementation used in this Chapter is much simpler mathematically and has the additional advantage of being able to employ existing *fast Fourier transform* (FFT) algorithms. This Fourier-Bessel method is the subject of the present chapter.

The starting point for this method is (4.9) and it's simplified version, (4.10). In this case, f_u is expanded in terms of a Fourier series

$$f_u(x, y) = \sum_{m,n} E_{mn}^u \, e^{-j\frac{2\pi}{D}(mx + ny)} \tag{5.1}$$

where D is the diameter of the projected aperture of the reflector. A Taylor-series expansion of the exponential (as in Chapter 4) gives

$$I_u = \sum_{p=0}^{P} \frac{1}{p!} \left(jk\frac{a^2}{4F} \right)^p (w - w_0)^p$$
$$\cdot \sum_{m,n} E_{mn}^u \int_0^1 \int_0^{2\pi} (s^2 - 1)^p \, e^{jk(\bar{u}x + \bar{v}y)} e^{-jk(\lambda/D)(mx + ny)} \, s \, ds \, d\phi' \tag{5.2}$$

We can combine the exponentials in the integrand to produce

$$I_u = \sum_{p=0}^{P} \frac{1}{p!} \left(jk\frac{a^2}{4F} \right)^p (w - w_0)^p$$

$$\cdot \sum_{m,n} E_{mn}^u \int_0^1 \int_0^{2\pi} (s^2 - 1)^p \, e^{jk(\bar{u}_m x + \bar{v}_n y)} \, s \, ds \, d\phi' \tag{5.3}$$

where

$$\bar{u}_m = \bar{u} - m\frac{\lambda}{D} \tag{5.4a}$$

$$\bar{v}_n = \bar{v} - n\frac{\lambda}{D} \tag{5.4b}$$

This last equation can be once again rewritten as

$$I_u = \sum_{p=0}^{P} \frac{1}{p!} \left(jk\frac{a^2}{4F} \right)^p (w - w_0)^p$$

$$\cdot \sum_{m,n} E_{mn}^u \int_0^1 \int_0^{2\pi} (s^2 - 1)^p \, e^{jkas\eta_{mn}\cos(\phi' - \alpha_{mn})} \, s \, ds \, d\phi' \tag{5.5}$$

where

$$\bar{u}_m = \eta_{mn} \cos \alpha_{mn} \tag{5.6a}$$

$$\bar{v}_n = \eta_{mn} \sin \alpha_{mn} \tag{5.6b}$$

and

$$x = as \cos \phi' \tag{5.7a}$$

$$y = as \sin \phi' \tag{5.7b}$$

At this point, it is now necessary to evaluate the integrals appearing in (5.5). These integrals are of the general form

$$I_p = \int_0^1 \int_0^{2\pi} (s^2 - 1)^p \, e^{jAs\cos\phi} \, s \, ds \, d\phi \tag{5.8}$$

Performing the ϕ-integration yields

$$I_p = 2\pi \int_0^1 (s^2 - 1)^p \, J_0(As) \, s \, ds \tag{5.9}$$

where $J_0(\cdot)$ is the Bessel function of order zero. If this integral is evaluated for $p = 0, 1, 2, 3$ using integration by parts, it is evident that

$$I_p = 2\pi\,(-1)^p\,2^p\,p!\,\frac{J_{p+1}(A)}{A^{p+1}} \tag{5.10}$$

where $J_{p+1}(\cdot)$ is the Bessel function of order $p + 1$. Hence,

$$I_u = 2\pi \sum_{p=0}^{P} \left(-jk\,\frac{a^2}{2F}\right)^p (w - w_0)^p \sum_{m,n} E_{mn}^u \frac{J_{p+1}(ka\eta_{mn})}{(ka\eta_{mn})^{p+1}} \tag{5.11}$$

The equation for the far zone radiated electric field then becomes

$$E_u(r) = -j\,\frac{\eta}{2}\,(ka^2)\,\frac{e^{-jkR}}{R}\,e^{jkd_{\text{off}}u}\,e^{jk\left(\frac{d_{\text{off}}^2 - 4F^2}{4F}\right)w}$$

$$\cdot\,e^{jk\frac{a^2}{4F}w} \sum_{m,n} E_{mn}^u \left\{\sum_{p=0}^{P} \left(-jk\,\frac{a^2}{2F}\right)^p (w - w_0)^p \frac{J_{p+1}(ka\eta_{mn})}{(ka\eta_{mn})^{p+1}}\right\} \tag{5.12}$$

where subtraction of the radial component is assumed. The coefficients E_{mn}^u ($u = x, y, z$) from (5.1) are calculated as

$$E_{mn}^u = \frac{1}{D^2} \int_{d_{\text{off}}-D}^{d_{\text{off}}+D} dx \int_{-D}^{D} dy\, f_u(x, y)\, e^{j\frac{2\pi}{D}(Mx + Ny)} \tag{5.13}$$

The integrand $f_u(x, y)$ in (5.13) is considered to extend over the entire $D \times D$ square area even though, in fact, it only extends over the circular projected aperture of diameter D inscribed within the $D \times D$ square.

Equations (5.12) and (5.13), along with the equations defining $f_u(x, y)$, are the final equations in the Fourier-Bessel method. The mathematics involved in this method are far less complex than those used in the Jacobi-Bessel method, making this a very practical analytical tool for most antenna engineers. The true beauty of this method however, lies in the equation for the Fourier coefficients of the aperture field distribution (5.13). Since this is an equation for coefficients of a Fourier series, it can be evaluated using readily available FFT software.

Another advantageous feature of the Fourier-Bessel method is that it is easily adapted to non-parabolic (i.e., shaped or distorted) reflectors. Due to the remarkable speed of the FFT algorithm, it is not critical to expand the exponential "dish depth" term as a Taylor series. We can merely lump this term into the effective aperture distribution f_u (x, y), and evaluate the Fourier coefficients E^u_{mn} again for each far-field direction. The FFT algorithm is so fast that we can do this and not suffer excessive loss in computational efficiency.

To illustrate the application of the Fourier-Bessel method to distorted reflectors, consider the problem of calculating the focal region fields of a distorted parabolic reflector illuminated by a plane electromagnetic wave. A schematic of this problem is shown in Figure 5.1. The focal region field of the distorted paraboloid is

$$\mathbf{E}(\mathbf{r}) = -j\frac{k\eta}{4\pi} \iint\limits_{\text{refl}} (\mathbf{I} - \hat{R}\hat{R}) \cdot \mathbf{J}(\mathbf{r}') \, G(\mathbf{r}|\mathbf{r}') \, ds' \tag{5.14}$$

where

$$\mathbf{r} = \text{a point in the focal region}$$

$$\mathbf{r}' = \text{a point on the reflector surface}$$

$$\mathbf{J}(\mathbf{r}') = 2\,\hat{n} \times \mathbf{H}^{\text{inc}}\, e^{jk\mathbf{r}'\cdot\hat{R}_{\text{inc}}}$$

$$G(\mathbf{r}|\mathbf{r}') = \frac{e^{-jk|\mathbf{r}-\mathbf{r}'|}}{|\mathbf{r} - \mathbf{r}'|} = \frac{e^{-jk\rho}}{\rho}$$

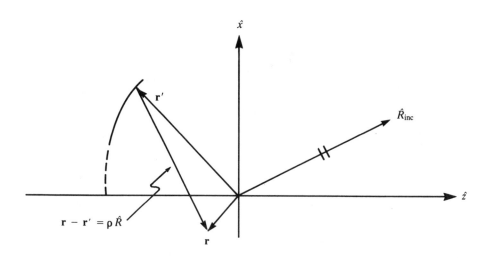

Figure 5.1 On the calculation of focal region fields of an offset reflector.

In this case, the operator $(\mathbf{I} - \hat{R}\hat{R})$ has been moved inside the integral sign to account for the fact that the focal region point is in the far field of each incremental area element on the reflector, but not in the far field of the entire reflector itself.

The phase of the current is given by

$$\text{phase} = j\,(u_{\text{inc}}x' + v_{\text{inc}}y') + jz'\,w_{\text{inc}} \tag{5.15}$$

where

$$u_{\text{inc}} = k \sin \theta_{\text{inc}} \cos \phi_{\text{inc}} \tag{5.16a}$$
$$v_{\text{inc}} = k \sin \theta_{\text{inc}} \sin \phi_{\text{inc}} \tag{5.16b}$$
$$w_{\text{inc}} = k \cos \theta_{\text{inc}} \tag{5.16c}$$

Likewise,

$$\mathbf{J}(\mathbf{r}') = 2\,\hat{n} \times \mathbf{H}^{\text{inc}} = \frac{2}{\eta}\,\hat{n} \times (\mathbf{E}^{\text{inc}} \times \hat{R}^{\text{inc}}) \tag{5.17}$$

So, the equation for the focal region electric field is

$$\mathbf{E}(\mathbf{r}) = -j/\lambda \iint_{\text{refl}} (\mathbf{I} - \hat{R}\hat{R}) \cdot [\hat{n} \times (\mathbf{E}^{\text{inc}} \times \hat{R}^{\text{inc}})] \frac{e^{-jk\rho}}{\rho}$$
$$\cdot\, e^{jz'w_{\text{inc}}} J_{\Sigma}\, e^{j(u_{\text{inc}}x' + v_{\text{inc}}y')}\, dx'\, dy' \tag{5.18}$$

If we now denote the uth component of the quantity

$$(\mathbf{I} - \hat{R}\hat{R}) \cdot [\hat{n} \times (\mathbf{E}^{\text{inc}} \times \hat{R}^{\text{inc}})] \frac{e^{-jk\rho}}{\rho}\, e^{jz'w_{\text{inc}}} J_{\Sigma}$$

by $g_u(x', y')$ and then expand this quantity in terms of a Fourier series as

$$g_u(x', y') = \sum_{m,n} g_u^{mn}\, e^{-j\frac{2\pi}{D}(mx' + ny')} \tag{5.19}$$

the integral for $E(r)$ becomes

$$E_u(r) = -j/\lambda \sum_{m,n} g_u^{mn} \iint_{\substack{\text{proj} \\ \text{aper}}} e^{j(u_{\text{inc}}x' + v_{\text{inc}}y')} \cdot e^{-j\frac{2\pi}{D}(mx' + ny')}\, dx'\, dy' \tag{5.20}$$

where the integral is evaluated over the projected circular aperture of the reflector. Proceeding as before, (5.20) can be put into the following form:

$$E_u(r) = -jk\left(\frac{D}{2}\right)^2 \sum_{m,n} g_u^{mn} \frac{J_1\left(\eta_{mn}\frac{D}{2}\right)}{\left(\eta_{mn}\frac{D}{2}\right)} \tag{5.21}$$

where

$$\eta_{mn} = \sqrt{\left(u_{\text{inc}} - \frac{2m\pi}{D}\right)^2 + \left(v_{\text{inc}} - \frac{2n\pi}{D}\right)^2} \tag{5.22}$$

Equation (5.21) gives the focal region electric field in terms of the expansion coefficients, g, and the incident field directions u_{inc}, v_{inc}. In this case, the expansion coefficients themselves depend on the incident field direction, and hence they must be calculated anew for every incident field direction and for every different focal region point.

From the form of (5.21), it is tempting to interpret this equation as an expansion for the radiated electric field in terms of sampling functions

$$\frac{J_1(\eta_{mn}D/2)}{(\eta_{mn}D/2)}$$

It must be emphasized that this equation is only valid for one incidence direction and does not express the field as a function of incidence angle. Chapter 6 will examine methods for applying sampling techniques to the calculation of radiation patterns of reflector antennas.

Suggestions for Additional Reading

1. Mittra, R., W.L. Ko, and M.S. Sheshadri, "A Transform Technique for Computing the Radiation Pattern of Prime-Focal and Cassegrainian Reflector Antennas," *IEEE Trans. Antennas Propagat.*, Vol. AP-30, pp. 520–524, May 1982.
2. Hung, C.C. and R. Mittra, "Secondary Pattern and Focal Region Distribution of Reflector Antennas Under Wide-Angle Scanning," *IEEE Trans. Antennas Propagat.*, Vol. AP-31, pp. 756–763, September, 1983.

Chapter 6
Sampling Methods

This chapter presents a variety of approaches to reflector antenna analysis which all—in one way or another—seek to calculate the radiation pattern by interpolating on the radiated field at certain sparsely located points in space. These methods, however, do not employ ordinary interpolation techniques. Instead, they stem from the Whittaker-Shannon sampling theorem used in communication theory.

Before delving too deeply into sampling theory, perhaps it is best to first introduce this approach by considering one particular sampling scheme which is actually quite closely related to the Fourier-Bessel method of Chapter 5. We begin with (4.11) which is repeated here in slightly modified form

$$I = \frac{1}{a^2} \int_0^a \int_0^{2\pi} f(r,\, \phi') e^{jk\frac{(r^2-a^2)}{4F}(w-w_0)}$$
$$\cdot\, e^{jkr\eta\cos(\phi'-\alpha)} r\, dr\, d\phi'$$
(6.1)

As in the Fourier-Bessel method, we expand the effective aperture distribution, f, in terms of a double Fourier series

$$f(r,\, \phi') = \sum_{m,n} f_{mn}\, e^{-j\frac{2\pi}{D}(mx+ny)}$$
(6.2)

where

$$f_{mn} = \frac{1}{4a^2} \int_{-a}^{a} \int_{-a}^{a} f(r,\, \phi')\, e^{j\frac{2\pi}{d}(mx+ny)}\, dx\, dy$$
(6.3)

The approach used here is somewhat different from that used in Chapter 5 since we now regard $f(r, \phi')$ as being zero outside the circular projected aperture for the purposes of calculating the Fourier coefficients in (6.3). On the other hand, when we calculate the radiation integral (6.1), we will integrate over the entire $2a \times 2a$ square surrounding the circular projected aperture. This is exactly the opposite of the procedure in Chapter 5 where we assumed $f(r, \phi')$ to exist over the entire $2a \times 2a$ square for purposes of calculating the Fourier coefficients, but only integrated over the circular projected aperture to calculate the radiation integral.

We can now substitute (6.2) into (6.1) to obtain

$$
I = \frac{1}{a^2} \sum_{m,n} f_{mn} \int\int_{-a}^{a} e^{jk\frac{(x^2+y^2-a^2)}{4F}(w-w_0)}
$$

$$
\cdot\, e^{j\left[\left(k\bar{u}-\frac{2m\pi}{D}\right)x + \left(k\bar{v}-\frac{2n\pi}{D}\right)y\right]} dx\, dy
$$

(6.4)

where $\bar{u}$ and $\bar{v}$ are related to η and α as in (4.12). Equation (6.4) can be factored into the form

$$
I = \frac{1}{a^2} e^{-j\frac{ka^2}{4F}(w-w_0)} \sum_{m,n} f_{mn} \int_{-a}^{a} e^{jk\frac{(w-w_0)}{4F}x^2} e^{j\left(k\bar{u}-\frac{2m\pi}{D}\right)x} dx
$$

$$
\cdot \int_{-a}^{a} e^{jk\frac{(w-w_0)}{4F}y^2} e^{j\left(k\bar{v}-\frac{2n\pi}{D}\right)y} dy
$$

(6.5)

This last equation can be written in the form

$$
I = \frac{4}{a^2} e^{-j\frac{ka^2}{4F}(w-w_0)}
$$

$$
\cdot \sum_{m,n} f_{mn}\Phi\left[\frac{k}{F}(w-w_0), k\bar{u}-\frac{2m\pi}{D}\right] \Phi\left[\frac{k}{F}(w-w_0), k\bar{v}-\frac{2n\pi}{D}\right]
$$

(6.6)

where

$$
\Phi[A,B] = \int_{-a/2}^{a/2} e^{j(Au^2+2Bu)}\, du
$$

(6.7)

The Φ integrals can be evaluated easily enough as

$$\Phi[A,B] = e^{-jB^2/A} \int_{-a/2}^{a/2} e^{jA\left(u^2 + 2\frac{B}{A}u + \frac{B^2}{A^2}\right)} du$$

$$(6.8)$$

$$= e^{-jB^2/A} \int_{-a/2}^{a/2} e^{jA\left(u + \frac{B}{A}\right)^2} du$$

If we let $t = \sqrt{A}\left(u + \dfrac{B}{A}\right)$, (6.8) becomes

$$\Phi[A,B] = \frac{1}{\sqrt{A}} e^{-jB^2/A} \int_{\sqrt{A}[(B/A)-(a/2)]}^{\sqrt{A}[(B/A)+(a/2)]} e^{jt^2} dt \qquad (6.9)$$

This last integral is easily expressed in terms of Fresnel integrals.

Equation (6.6) is the key equation here. It expresses the radiation integral at a general point $\bar{u}, \bar{v}$, w in space in terms of the Fourier transforms of the aperture distribution at

$$\bar{u} = \frac{2m\pi}{kD}, \bar{v} = \frac{2n\pi}{kD} \qquad (6.10)$$

where D = diameter of the circular projected aperture of the reflector. Since $k = 2\pi/\lambda$, we have

$$\bar{u} = m\frac{\lambda}{D} \quad \bar{v} = n\frac{\lambda}{D} \qquad (6.11)$$

Equation (6.11) shows the sample density in $\bar{u}, \bar{v}$ space to be

$$\Delta\bar{u} = \Delta\bar{v} = \lambda/D \qquad (6.12)$$

That is, the samples are taken roughly at the rate of one per lobe.

The Φ functions in (6.6) are generalizations of the usual cardinal functions $\sin x/x$ used in Whittaker-Shannon sampling. As before in Chapter 5, the FFT can be used to calculate the Fourier coefficients, f_{mn}. The particular choice of Φ used here only applies for parabolic reflectors. As noted in [1], however, the Φ functions could be generated numerically for the case of shaped or distorted reflectors.

If the focal length, F, of the parabola becomes infinite (i.e., the parabolic reflector becomes planar), (6.5) becomes

$$I = \frac{1}{a^2} \sum_{m,n} f_{mn} \int_{-a}^{a} e^{j(k\bar{u} - 2m\pi/D)x} \, dx$$

$$\int_{-a}^{a} e^{j(k\bar{v} - 2n\pi/D)y} \, dy \qquad (6.13)$$

These integrals are immediately evaluated as

$$I = 4 \sum_{m,n} f_{mn} \, \text{sinc} \left(k\bar{u} - \frac{2m\pi}{D} \right) a \, \text{sinc} \left(k\bar{v} - \frac{2n\pi}{D} \right) a \qquad (6.14)$$

where sinc $x = \sin x / x$. Rearranging the sinc function arguments gives

$$I(\bar{u}, \bar{v}) = 4 \sum_{m,n} f_{mn} \, \text{sinc} \, \pi \left(\frac{D}{\lambda} \bar{u} - m \right) \text{sinc} \, \pi \left(\frac{D}{\lambda} \bar{v} - n \right) \qquad (6.15)$$

Equation (6.15) is significant because it expresses a continuous function $I(\bar{u}, \bar{v})$ in terms of a set of equally spaced samples f_{mn} of the Fourier transform of the aperture distribution. In fact,

$$I(\bar{u} = m\lambda/D, \bar{v} = n\lambda/D) = 4f_{mn} \qquad (6.16)$$

an equality that can be deduced immediately from the equation.

Substituting (6.16) into (6.15) yields

$$I(\bar{u}, \bar{v}) = \sum_{m,n} I_{mn} \, \text{sinc} \, \pi \left(\frac{D}{\lambda} \bar{u} - m \right) \text{sinc} \, \pi \left(\frac{D}{\lambda} \bar{v} - n \right) \qquad (6.17)$$

where

$$I_{mn} = I(m\lambda/D, n\lambda/D) \qquad (6.18)$$

This is a two dimensional version of the Whittaker-Shannon sampling theorem. Equation (6.17) essentially states that $I(\bar{u}, \bar{v})$ can be reconstructed as a continuous function of $\bar{u}, \bar{v}$ using only the values of I at the discrete set of sample points $\bar{u}_m$, $\bar{v}_n$.

Now that we have seen how the radiation integral can be rewritten in the form of a sampling series, it is worthwhile to study the Whittaker-Shannon sampling theorem more formally. This theorem may be stated (in one dimension) as follows: if the function $f(x)$ is non-zero only in the interval $-a \leqslant x \leqslant a$ (the interval may actually be arbitrary as long as it is finite), then the Fourier transform of $f(x)$, as defined by

$$f(x) = \frac{1}{2\pi} \int_{-\infty}^{\infty} F(k) \, e^{-jkx} \, dk \qquad (6.19)$$

$$F(k) = \int_{-\infty}^{\infty} f(x) \, e^{jkx} \, dx \qquad (6.20)$$

may be calculated exactly using the infinite series

$$F(k) = \sum_{n=-\infty}^{\infty} F(k_n) \, \text{sinc} \, \pi(k - k_n)/\Delta k \qquad (6.21)$$

where

$$k_n = n\Delta k \qquad (6.22)$$

and

$$\Delta k < \frac{\pi}{a} \qquad (6.23)$$

Before proving the theorem, we will digress briefly to show that (6.19) and (6.20) are actually Fourier transform pairs. Substituting (6.20) into (6.19) produces

$$f(x) = \frac{1}{2\pi} \int_{-\infty}^{\infty} \left\{ \int_{-\infty}^{\infty} f(x') \, e^{jk'x'} \, dx' \right\} e^{-jk'x} \, dk' \qquad (6.24)$$

Rearranging (6.24) gives

$$f(x) = \frac{1}{2\pi} \int_{-\infty}^{\infty} f(x') \, dx' \int_{-\infty}^{\infty} e^{jk'(x'-x)} \, dk' \qquad (6.25)$$

However,

$$\int_{-\infty}^{\infty} e^{j(\omega-\omega_0)t} \, dt = 2\pi \, \delta(\omega - \omega_0) \qquad (6.26)$$

by [2]. Hence (6.25) becomes

$$f(x) = \int_{-\infty}^{\infty} f(x') \, \delta(x - x') \, dx' = f(x) \qquad (6.27)$$

Equations (6.19) and (6.20) do indeed represent Fourier transform pairs.

The actual proof of the sampling theorem is quite easy. Because we specified $f(x)$ to be zero outside the interval $[-a, a]$, we can say that $f(x)$ has the general form shown in Figure 6.1. Since $f(x) = 0$ outside the interval $[-a, a]$, we can just as well say that $f(x)$ is equal to the periodic extension of $f(x)$ (shown in Figure 6.2) multiplied by the unit pulse function of width $2A$, centered at $x = 0$. From Figure 6.2, it is clear that as long as $A > a$, we can recover $f(x)$ from its periodic extension simply by multiplying $f_p(x)$ by the unit height pulse function. In other words,

$$f(x) = p_{2A}(x) f_p(x) \qquad (6.28)$$

as long as $A > a$.

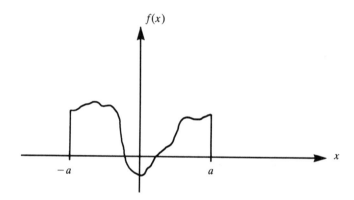

Figure 6.1 A general space-limited function, $f(x)$.

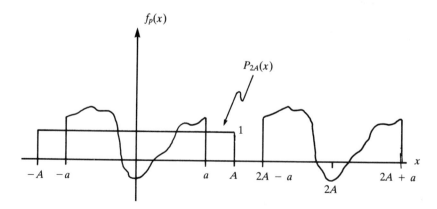

Figure 6.2 Periodic extension of $f(x)$.

Since $f_p(x)$ is periodic in x, with periodicity $2A$, it may be represented as a Fourier series of the form

$$f_p(x) = \sum_{n=-\infty}^{\infty} c_n e^{-j\frac{2n\pi}{2A}x} \qquad (6.29)$$

where

$$c_n = \frac{1}{2A} \int_{-A}^{A} f_p(x) e^{j\frac{n\pi}{A}x} dx \qquad (6.30)$$

Substituting (6.29) into (6.28) gives

$$f(x) = p_{2A}(x) \sum_{n=-\infty}^{\infty} c_n e^{-j\frac{n\pi}{A}x} \qquad (6.31)$$

Since $f(x) = 0$ for $|x| > a$, (6.30) becomes

$$c_n = \frac{1}{2A} \int_{-\infty}^{\infty} f(x) e^{j\frac{n\pi}{A}x} dx \qquad (6.32)$$

or

$$c_n = \frac{1}{2A} F\left(\frac{n\pi}{A}\right) \qquad (6.33)$$

Substituting (6.33) into (6.31) yields

$$f(x) = p_{2A}(x) \sum_{n=-\infty}^{\infty} \frac{1}{2A} F\left(\frac{n\pi}{A}\right) e^{-j\frac{n\pi}{A}x} \qquad (6.34)$$

Now we merely Fourier transform (6.34) term by term according to (6.20) to give

$$F(k) = \sum_{n=-\infty}^{\infty} F\left(\frac{n\pi}{A}\right) \text{sinc}\,(kA - n\pi) \qquad (6.35)$$

The integral of the term $\exp[j(k - n\pi/A)x]$ over the interval $[-A, A]$ produces the term 2A sinc $(kA - n\pi)$. We see from (6.35) that

$$\Delta k = \frac{\pi}{A} \text{ and } k_n = n\frac{\pi}{A} = n\,\Delta k \tag{6.36}$$

So, the argument of the sinc function is

$$kA - n\pi = A\left(k - \frac{n\pi}{A}\right) = A(k - k_n)$$
$$= \pi(k - k_n)/\Delta k \tag{6.37}$$

as in (6.21). Also, our condition $A > a$ is equivalent to

$$\Delta k = \frac{\pi}{A} < \frac{\pi}{a} \tag{6.38}$$

as in (6.23). The spatial frequency $k = \pi/a$ is the coarsest sampling permissible in order for the reconstruction of $F(k)$ in (6.35) to be exact. This frequency is often referred to as the Nyquist frequency.

We can directly apply the sampling theorem to the problem of calculating the far-zone radiation pattern of a reflector antenna [3]. To do this, consider Figure 6.3, which shows the problem schematically. If it were the case that the electric field radiated by the reflector in Figure 6.3 was non-zero only over some finite portion of the plane P, then we could use the sampling theorem to calculate the far-zone electric field (which is merely the Fourier transform of the electric field on the plane, P). Clearly, the closer P resides to the rim of the reflector, the faster the electric field on P will decay as we move away from the projection of the aperture onto P. However, even when P lies right up against the uppermost point

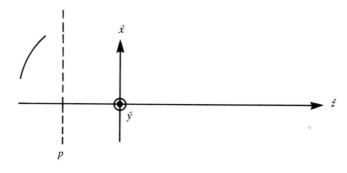

Figure 6.3 Schematic of offset reflector.

of this offset reflector, the electric field on P will never be space-limited. On the other hand, for the case of a focused reflector, we expect the fields on P to decay rapidly away from the projected aperture.

Therefore, even though the fields on P are not strictly space-limited, it appears likely that they will be space-limited enough so that we can use the sampling theorem to calculate far fields. The transverse (x, y component) fields $E_P(x, y)$ and $E_P(\bar{u}, \bar{v})$ are Fourier transform pairs. We can say by the sampling theorem that

$$E_P(\bar{u}, \bar{v}) = \sum_{m,n} E_P(\bar{u}_m, \bar{v}_n) \, \text{sinc} \, \pi(\bar{u} - \bar{u}_m)/\Delta u \, \text{sinc} \, \pi(\bar{v} - \bar{v}_n)/\Delta v \quad (6.39)$$

The actual far-zone radiated electric fields are related to the Fourier transform of the transverse field components on P by the relation

$$\mathbf{E}_{\text{rad}}(\bar{u}, \bar{v}) = (\mathbf{I} - \hat{R}\hat{R}) \cdot \mathbf{E}_P(\bar{u}, \bar{v}) \quad (6.40)$$

Substituting (6.39) into (6.40) yields

$$\mathbf{E}_{\text{rad}}(\bar{u}, \bar{v}) = (\mathbf{I} - \hat{R}\hat{R}) \cdot \sum_{m,n} (\mathbf{I} - \hat{R}\hat{R})^{-1} \mathbf{E}_{\text{rad}}(\bar{u}_m, \bar{v}_n)$$
$$\cdot \, \text{sinc} \, \pi(\bar{u} - \bar{u}_m)/\Delta u \, \text{sinc} \, \pi(\bar{v} - \bar{v}_n)/\Delta v \quad (6.41)$$

So, this is a sampling type expression pertaining to radiated far fields of the reflector.

The following comments are in order. The radiated far field at the sample points $\bar{u}_m$, $\bar{v}_n$ may be calculated using ordinary means (i.e., by any of the methods in the previous chapters), since the sample points are relatively sparse in $\bar{u}$, $\bar{v}$ space. The radiation pattern is then "filled in" using the sampling expansion (6.41).

If the electric field on P were perfectly space-limited to the aperture projected area, then the coarsest sampling allowed by the sampling theorem would be

$$\Delta \bar{u} = \Delta \bar{v} = \frac{\lambda}{2a} \quad (6.42)$$

where a is the radius of the reflector projected aperture onto the x, y plane.

Note: (6.42) is the same as (6.12).

Finally, we observe that the operator $\mathbf{I} - \hat{R}\hat{R}$ may be represented by the following matrix:

$$
\begin{bmatrix} E_\theta \\ E_\phi \end{bmatrix} = \begin{bmatrix} \cos\theta\cos\phi & \cos\theta\sin\phi \\ -\sin\phi & \cos\phi \end{bmatrix} \begin{bmatrix} E_x \\ E_y \end{bmatrix}
\tag{6.43}
$$

or

$$
\begin{bmatrix} E_x \\ E_y \end{bmatrix} = \begin{bmatrix} \cos\phi/\cos\theta & -\sin\phi \\ \sin\phi/\cos\theta & \cos\phi \end{bmatrix} \begin{bmatrix} E_\theta \\ E_\phi \end{bmatrix}
\tag{6.44}
$$

It is also possible to employ non-uniform sampling techniques for calculating radiation patterns [4]. Two possible approaches are illustrated below. As shown in the derivation of the sampling theorem, the Fourier transform of a space-limited function may be expressed by the series

$$
F(k) = \sum_{n=1}^{N} F(k_n)\, \psi_n(k)
\tag{6.45}
$$

where

$$
\psi_n(k) = \operatorname{sinc} \pi(k - k_n)/\Delta k
\tag{6.46}
$$

and

$$
k_n = n\Delta k
\tag{6.47}
$$

In the event that the samples are non-uniform in k, we can still use (6.45) to calculate $F(k)$, except that we now write $\psi_n(k)$ as

$$
\psi_n(k) = \sum_{p=1}^{N} A_{np} \operatorname{sinc} \pi(k - k_p)/\Delta k
\tag{6.48}
$$

where the coefficients A_{np} are yet to be determined. Note that in (6.48), $k_p \neq p\,\Delta k$. However, we still require that

$$
\Delta k < \frac{\pi}{a}
\tag{6.49}
$$

as before. Also, it is not advisable to allow spacings between any two adjacent points to exceed the criterion in (6.49), even though the sample points are not uniformly spaced.

The new sampling function $\psi_n(k)$ in (6.48) no longer sifts out the values of $F(k)$ at the sample points k_n. For example, A_{np} is no longer a Kronecker delta function. So, to solve for the coefficients A_{np}, we enforce the following conditions:

$$F(k_m) = \sum_{n=1}^{N} F(k_n) \left[\sum_{p=1}^{P} A_{np} \text{ sinc } \pi(k_m - k_p)/\Delta k \right] \tag{6.50}$$

for $m = 1, 2, \ldots, N$. We obtain (6.50) by substituting (6.48) into (6.45) and evaluating at the sample points, k_m. In order to satisfy (6.50), we must have

$$\sum_{p=1}^{P} A_{np} \text{ sinc } \pi(k_m - k_p)/\Delta k = \delta_{mn}, (m = 1, 2, \ldots, N) \tag{6.51}$$

where, as before, δ_{mn} is the Kronecker delta function defined by

$$\delta_{mn} = 1 \quad (m = n) \tag{6.52}$$
$$= 0 \quad (m \neq n)$$

If we denote

$$C_{pm} = \text{sinc } \pi(k_m - k_p)/\Delta k \tag{6.53}$$

then (6.51) becomes

$$\sum_{p=1}^{P} A_{np} C_{pm} = \delta_{mn} \tag{6.54}$$

or equivalently,

$$[A][C] = [I] \tag{6.55}$$

in matrix form. Equation (6.55) gives

$$[A] = [C]^{-1} \tag{6.56}$$

This is our first method for extending the sampling theorem to non-uniformly spaced sample points. In the second method, we use the original interpolation function (i.e., sinc $(\cdot)$) and calculate a new set of coefficients. For example

$$F(k) = \sum_{n=1}^{N} A_n \text{ sinc } \pi(k - k_n)/\Delta k \qquad (6.57)$$

The coefficients A_n are easily determined by setting

$$F(k_p) = \sum_{n=1}^{N} A_n \text{ sinc } \pi(k_p - k_n)/\Delta k, \ (p = 1, 2, \ldots, N) \qquad (6.58)$$

Rewriting (6.58) in matrix form gives

$$[C] \, A = F \qquad (6.59)$$

where

$$C_{mn} = \text{sinc } \pi(k_m - k_n)/\Delta k \qquad (6.60)$$

$$F_m = F(k_m) \qquad (6.61)$$

This concludes our discussion on sampling methods.

REFERENCES

1. Bucci, O.M., G. Franceschetti, and R. Pierri, "Reflector Antenna Fields—An Exact Aperture-Like Approach," *IEEE Trans. Antennas Propagat.*, Vol. AP-29, pp. 580–586, July 1981.
2. Papoulis, A., *Systems and Transforms with Applications in Optics*, Krieger: Malabar, FL, 1981.
3. Bucci, O.M., G. Franceschetti, and G. D'Elia, "Fast Analysis of Large Antennas—A New Computational Philosophy," *IEEE Trans. Antennas Propagat.*, Vol. AP-28, pp. 306–310, May 1980.
4. Rahmat-Samii, Y. and R.L.T. Cheung, "Nonuniform Sampling Techniques for Antenna Applications," *IEEE Trans. Antennas Propagat.*, Vol. AP-35, pp. 268-279, March 1987.

Suggestions for Additional Reading

1. Bucci, O.M., G. D'Elia, G. Franceschetti, and R. Pierri, "Efficient Computation of the Far Field of Parabolic Reflectors by Pseudo-Sampling Algorithm," *IEEE Trans. Antennas Propagat.*, Vol. AP-31, pp. 931–937, November 1983.
2. Bucci, O.M. and G. Di Massa, "Exact Sampling Approach for Reflector Antennas Analysis," *IEEE Trans. Antennas Propagat.*, Vol. AP-32, pp. 1259–1262, November 1984.
3. Bucci, O.M. and G. Franceschetti, "On the Spatial Bandwidth of Scattered Fields," *IEEE Trans. Antennas Propagat.*, Vol. AP-35, pp. 1445–1455, December 1987.
4. Yen, J.L., "On Nonuniform Sampling of Bandwidth—Limited Signals," *IRE Trans. Circuit Theory*, pp. 251–257, December 1956.

Chapter 7
The Quadratic Phase Method

We will conclude the discussion on reflector antenna analysis techniques with a method which represents a generalization of the Ludwig algorithm from Chapter 2. In this approach, the rapidly varying phase terms are approximated by quadratics rather than linear functions [1–5]. This method is surprisingly flexible and has been generalized to allow for arbitrary polynomials in both the magnitude and phase portions of the integrand [6]. On the other hand, the quadratic phase algorithm is basically a one dimensional scheme, whereas the Ludwig algorithm is two dimensional.

To illustrate this method, consider the radiation integral from (4.11), which is rewritten in (7.1):

$$I = \int_0^1 \int_0^{2\pi} f(s, \phi') e^{jk\frac{a^2}{4F}(s^2-1)(w-w_0)} \, e^{jkas\eta\cos(\phi'-\alpha)} \, s \, ds \, d\phi' \tag{7.1}$$

The effective aperture distribution $f(s, \phi')$ is expanded in Fourier series form as

$$f(s, \phi') = \sum_n a_n(s) \cos n\phi' + b_n(s) \sin n\phi' \tag{7.2}$$

Hence

$$
\begin{aligned}
I = {} & 2\pi \sum_n j^n \cos n\alpha \int_0^1 a_n(s) \, e^{jk\frac{a^2}{4F}(s^2-1)(w-w_0)} \, J_n(kas\eta) \, s \, ds \\
& + 2\pi \sum_n j^n \sin n\alpha \int_0^1 b_n(s) \, e^{jk\frac{a^2}{4F}(s^2-1)(w-w_0)} J_n(kas\eta) \, s \, ds
\end{aligned}
\tag{7.3}
$$

The radial integral is now integrated numerically, by dividing up the reflector projected aperture into several concentric circular annuli and then integrating numerically over each annulus. The integrands in (7.3) have phase terms which are quadratic in s. The "magnitude" terms vary as

$$\begin{Bmatrix} a_n(\theta) \\ b_n(\theta) \end{Bmatrix} J_n(kas\eta)$$

A fundamental property of the Bessel function, $J_n(x)$, is that it varies approximately as

$$\frac{1}{n!}\left(\frac{x}{2}\right)^n \quad (x < n)$$

and as

$$\sqrt{\frac{2}{\pi x}} \cos\left(x - \frac{n\pi}{2} - \frac{\pi}{4}\right) \quad (x > n)$$

Because $J_n(x)$ is oscillatory for $x > n$, it makes sense to combine that oscillatory behavior with the quadratic exponential in the integrand. For $x < n$, we simply combine the Bessel function with the Fourier coefficients of the aperture distribution. So, we seek to evaluate an integral of the form

$$\begin{aligned} I &= \int_0^1 a_n(s) J_n(kas\eta) e^{jAs^2} ds \\ &= \int_0^\sigma a_n(s) J_n(kas\eta) e^{jAs^2} ds \\ &\quad + \int_\sigma^1 a_n(s) J_n(kas\eta) e^{jAs^2} ds \end{aligned} \tag{7.4}$$

where $\sigma = n/ka\eta$, and the $a_n(s)$ may be complex.

We will consider the two integrals separately. In the first case, where $0 \leqslant s \leqslant \sigma$, the Bessel function has the form

$$J_n(x) = \frac{1}{n!}\left(\frac{x}{2}\right)^n \cdot (\text{polynomial in } x) \tag{7.5}$$

Also,

$$a_n(s) = \text{polynomial in } s \tag{7.6}$$

So,

$$a_n(s) J_n(kas\eta) = \frac{1}{n!}\left(\frac{x}{2}\right)^n \cdot (\text{polynomial in } s) \tag{7.7}$$

For example, the first n terms in the polynomial expansion for

$$a_n(s) J_n (kas\eta)$$

are zero. This is worth knowing because it helps to avoid taking unnecessary terms in the expansion. The first integral in (7.4) becomes

$$I_1 = \int_0^\sigma s^n \, p(s) \, e^{jAs^2} \, ds \tag{7.8}$$

It is also worth noting that

$$J_n(0) = \delta_n \tag{7.9}$$

where δ_n is the Kronecker delta.

In the case of the integral in (7.8), the quadratic exponent is known explicitly, i.e., no quadratic approximation is required. Only the polynomial $p(s)$ needs to be approximated. This is done easily using a variety of standard techniques. Therefore, the key to evaluating (7.8) is to be able to evaluate the constituent integrals of the form

$$I_n = \int_{s_1}^{s_2} s^n \, e^{jAs^2} \, ds \tag{7.10}$$

These are easily evaluated using the product rule

$$\begin{aligned} I_n &= \frac{1}{2jA} \int_{s_1}^{s_2} \left(2jAs \, e^{jAs^2}\right) s^{n-1} \, ds \\ &= \frac{1}{2jA}\left\{ s^{n-1} \, e^{jAs^2}\Big|_{s_1}^{s_2} - (n-1)\int_{s_1}^{s_2} s^{n-2} \, e^{jAs^2} \, ds \right\} \end{aligned} \tag{7.11}$$

So,

$$I_n = \frac{1}{2jA}\left\{ s^{n-1} \, e^{jAs^2}\Big|_{s_1}^{s_2} - (n-1) I_{n-2} \right\} \tag{7.12}$$

This gives a recursion relation for the I_n. It can be rewritten as

$$I_{n-2} = \frac{1}{n-1}\left\{ 2jA\ I_n - s^n\ e^{jAs^2}\Big|_{s_1}^{s_2}\right\} \tag{7.13}$$

Pogorzelski [2] has shown that this recursion can be solved by assuming $I_N = 0$ for some large value of N, and then recursing backward to calculate I_n for all $n < N$. Since the recursion only involves every other integer, a separate recursion must be performed for both even and odd integers, n.

The integral from σ to 1 can be handled similarly. From the asymptotic form of the Bessel function, we see that

$$J_n(x) \cong \sqrt{\frac{2}{\pi x}} \cdot \frac{1}{2}\left[e^{j\left(x - \frac{n\pi}{2} - \frac{\pi}{4}\right)} + e^{-j\left(x - \frac{n\pi}{2} - \frac{\pi}{4}\right)}\right] \tag{7.14}$$

which we approximate as

$$J_n(x) \cong p(x)\left[e^{j\left(x - \frac{n\pi}{2} - \frac{\pi}{4}\right)} + e^{-j\left(x - \frac{n\pi}{2} - \frac{\pi}{4}\right)}\right] \tag{7.15}$$

The linear phase variation in the Bessel function may be combined with the quadratic phase variation in the integrand, isolating the rapidly oscillatory portions of the integrand. From here, the solution proceeds as before.

Other variations on this quadratic phase scheme are possible. For example, the polynomial part of the integrand in (7.8) may be expanded in terms of Chebyschev polynomials [5]. In this case, we will denote the polynomial portion of (7.8) as

$$f(s) = \sum_{n=0}^{N} a_n T_n(s) \tag{7.16}$$

where

$$f(s) = s^n p(s) \tag{7.17}$$

The Chebyschev polynomials are defined by the relations

$$T_n(\cos \theta) = \cos n\theta \tag{7.18}$$

$$U_n(\cos \theta) = \frac{\sin (n + 1)\theta}{\sin \theta} \tag{7.19}$$

and they are polynomials in the variable $x = \cos \theta$.

First it is necessary to calculate the coefficients a_n in (7.16). To do this, we begin by defining a new variable, u, according to the relation

$$u = 2\left(\frac{s - s_1}{s_2 - s_1}\right) - 1 \tag{7.20}$$

or

$$s = s_1 + (s_2 - s_1)\left(\frac{u + 1}{2}\right) \tag{7.21}$$

So the variable, s, ranges from s_1 to s_2 and u ranges from -1 to $+1$. Now, (7.16) becomes

$$f(u) = \sum_{n=0}^{N} a_n T_n(u) \tag{7.22}$$

which may be rewritten directly as

$$f(\cos \theta) = \sum_{n=0}^{N} a_n \cos n\theta \tag{7.23}$$

where

$$u = \cos \theta$$

Expressing f directly in terms of the variable θ gives

$$f(\theta) = \sum_{n=0}^{N} a_n \cos n\theta \tag{7.24}$$

from which it is clear that this is a Fourier cosine series representation for the function $f(\theta)$, which is an even function of θ. The coefficients may be evaluated using an FFT or, if there are not too many terms in the series, by merely multiplying each side by $\cos m\theta$ and integrating over one period of the function, as in ordinary Fourier analysis.

Once the expansion coefficients, a_n, are known, the constituent integrals

$$I_n = \int_{-1}^{1} T_n(u) \, e^{jAu^2} \, du \tag{7.25}$$

need to be evaluated. As before, our strategy will be to determine a recursive procedure for evaluating these integrals. We begin with the following recursion relation for the Chebyschev polynomials

$$x \, T_n(x) = \frac{1}{2} T_{n-1}(x) + \frac{1}{2} T_{n+1}(x) \tag{7.26}$$

which is merely a disguised form of the trigonometric identity

$$2 \cos \theta \cos n\theta = \cos (n+1)\theta + \cos (n-1)\theta \tag{7.27}$$

Now integrate (7.26) from -1 to $+1$ against the quadratic phase term, $\exp (jAx^2)$, to get

$$\int_{-1}^{1} T_n(x) \, x \, e^{jAx^2} \, dx = \frac{1}{2} \int_{-1}^{1} T_{n-1}(x) \, e^{jAx^2} \, dx$$

$$+ \frac{1}{2} \int_{-1}^{1} T_{n+1}(x) \, e^{jAx^2} \, dx \tag{7.28}$$

$$= \frac{1}{2} I_{n-1} + \frac{1}{2} I_{n+1}$$

The left-hand side of (7.28) can be integrated by parts to give

$$\int_{-1}^{1} T_n(x)[x \, e^{jAx^2}] \, dx = T_n(x) \left. \frac{e^{jAx^2}}{2jA} \right|_{-1}^{1}$$

$$- \frac{1}{2jA} \int_{-1}^{1} T_n'(x) \, e^{jAx^2} dx \tag{7.29}$$

Equating (7.28) and (7.29) gives

$$\frac{T_n(x) \, e^{jAx^2}}{2jA} \bigg|_{-1}^{1} - \frac{1}{2jA} \int_{-1}^{1} T_n'(x) \, e^{jAx^2} \, dx$$

$$= \frac{1}{2}(I_{n-1} + I_{n+1}) \tag{7.30}$$

We now employ the following trigonometric identity

$$T_n'(x) = n \, U_{n-1}(x) \tag{7.31}$$

which is obtained by differentiating

$$f(x) = \cos n\theta \tag{7.32}$$

by

$$x = \cos \theta \tag{7.33}$$

which gives

$$\frac{d}{dx} \cos n\theta = \frac{d\theta}{dx} \frac{d}{d\theta} \cos n\theta$$

$$= \frac{-n \sin n\theta}{-\sin \theta} \tag{7.34}$$

$$= n \, U_{n-1}(x)$$

Having established the identity (7.31), we now integrate it against $\exp(jAx^2)$ from -1 to $+1$ to give

$$\int_{-1}^{1} T_n'(x) \, e^{jAx^2} \, dx = n \int_{-1}^{1} U_{n-1}(x) \, e^{jAx^2} \, dx$$

$$= n \, K_{n-1} \tag{7.35}$$

Therefore, (7.30) becomes

$$\frac{T_n(x) \, e^{jAx^2}}{2jA} \bigg|_{-1}^{1} - \frac{1}{2jA} \cdot n \, K_{n-1} = \frac{1}{2} I_{n-1} + \frac{1}{2} I_{n+1} \tag{7.36}$$

One final trigonometric identity is invoked to produce a recursion on the K_n. This is,

$$U_n(x) = 2 \, T_n(x) + U_{n-2}(x) \tag{7.37}$$

which is obtained from the trigonometric relation

$$\sin(n+1)\theta - \sin(n-1)\theta = 2 \cos n\theta \sin \theta \tag{7.38}$$

We rewrite (7.37) as

$$T_n(x) = \frac{1}{2}U_n(x) - \frac{1}{2}U_{n-2}(x) \tag{7.39}$$

and integrate against exp (jAx^2) from -1 to $+1$ to get

$$I_n = \frac{1}{2}K_n - \frac{1}{2}K_{n-2} \tag{7.40}$$

Substituting (7.40) into (7.36) gives

$$\frac{T_n(x)\,e^{jAx^2}}{2jA}\bigg|_{-1}^{1} = \frac{n}{2jA}K_{n-1} + \frac{1}{2}\left(\frac{1}{2}K_{n-1} - \frac{1}{2}K_{n-3}\right) + \frac{1}{2}\left(\frac{1}{2}K_{n+1} - \frac{1}{2}K_{n-1}\right) \tag{7.41}$$

This is the desired recursion on the $K_n's$. Once the $K_n's$ are known, the I_n can be solved for using (7.40).

As mentioned in [5], this recursion may be solved using a lower/upper triangular matrix decomposition procedure using assumed asymptotic values of

$$K_n = \frac{1}{n+1}\,T_{n+1}(x)\,e^{jAx^2}\bigg|_{-1}^{1} \tag{7.42}$$

One additional variation on this procedure involves generalizing the quadratic phase term to a polynomial of arbitrary degree. The derivation of the recursion proceeds along similar lines as that for the quadratic phase case and it may be referenced in the literature [6].

Note: The sample point density required may depend not only on the variation of the phase of the integrand, but also on obtaining an adequate definition of the surface or the feed pattern.

REFERENCES

1. Pogorzelski, R.J., "A New Integration Algorithm and its Application to the Analysis of Symmetrical Cassegrain Microwave Antennas," *IEEE Trans. Antennas Propagat.*, Vol. AP-31, pp. 748–755, September 1983.
2. Pogorzelski, R.J., "Comments on 'A New Integration Algorithm and its Application to the Analysis of Symmetrical Cassegrain Microwave Antennas'," *IEEE Trans. Antennas Propagat.*, Vol. AP-32, p. 431, April 1984.

3.   Pogorzelski, R.J., "Correction to 'A New Integration Algorithm and its Application to the Analysis of Symmetrical Cassegrain Microwave Antennas'," *IEEE Trans. Antennas Propagat.*, Vol. AP-32, p. 655, May 1984.

4.   Pogorzelski, R.J., "On the Numerical Analysis of the Subreflector of an Offset Cassegrain Microwave Antenna," *IEEE Trans. Antennas Propagat.*, Vol. AP-32, pp. 595–601, June 1984.

5.   Pogorzelski, R.J., "Quadratic Phase Integration Using a Chebyschev Expansion," *IEEE Trans. Antennas Propagat.*, Vol. AP-33, pp. 563–566, May 1985.

6.   Pogorzelski, R.J., "Complex Polynomial Phase Integration," *IEEE Trans. Antennas Propagat.*, Vol. AP-33, pp. 800–805, July 1985.

Chapter 8
Design of Single-Reflector Antenna Systems

The balance of this text will be devoted to various aspects of reflector antenna design. The subject of antenna design is not nearly as well defined as antenna analysis, so the presentation will not be as unified as before. The primary goal here is to alert the reader to the various kinds of practical issues that arise in reflector antenna design and to indicate effective ways of handling them.

8.1 APERTURE ILLUMINATION AND FEED SPILLOVER

We can begin by considering Figure 8.1, a schematic diagram of a symmetrical front-fed parabolic reflector antenna. It is of interest to know what type of feed pattern will uniformly illuminate the reflector projected aperture. Suppose the feed is located at the focal point of the paraboloid and has a radiation pattern of the form $e_f^2(\theta)$ watts/steradian. Now,

$$\rho = \frac{2F}{1 + \cos \theta} \tag{8.1}$$

$$r = 2F \frac{\sin \theta}{1 + \cos \theta} \tag{8.2}$$

and

$$dr = F \sec^2 \frac{\theta}{2} d\theta \tag{8.3}$$

Hence,

$$r \, dr \, d\phi = F^2 \sec^4 \frac{\theta}{2} \sin \theta \, d\theta \, d\phi \tag{8.4}$$

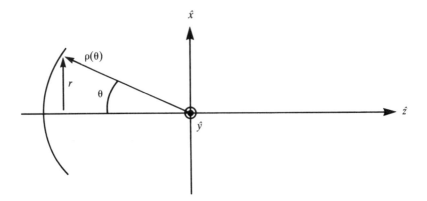

Figure 8.1 Symmetrical front-fed parabolic reflector.

Equation (8.4) can be rewritten as

$$da = F^2 \sec^4 \frac{\theta}{2} \ d\Omega \tag{8.5}$$

where

da = area element on projected aperture
$d\Omega$ = incremental solid angle

The total power contained in a ray tube extending from the feed to the reflector and then to the projected aperture is

$$e_f^2(\theta) \ d\Omega = e_f^2(r) \ da \tag{8.6}$$

We want $e_f^2(r) = 1$, hence

$$e_f^2(\theta) = \frac{da}{d\Omega} = \sec^4 \frac{\theta}{2} \tag{8.7}$$

where any constant factors have been neglected. So,

$$|e_f(\theta)| = \sec^2 \frac{\theta}{2} \tag{8.8}$$

As mentioned at the end of Chapter 3, the ideal vector feed pattern would be of the form

$$|e_\theta (\theta)| = \sec^2 \frac{\theta}{2} \cos \phi \tag{8.9}$$

$$|e_\phi (\theta)| = \sec^2 \frac{\theta}{2} \sin \phi \tag{8.10}$$

This ideal *secant squared* feed pattern is shown in Figure 8.2. A feed antenna having this ideal pattern would yield an antenna with a uniformly illuminated aperture and no spillover loss. While no feed would be able to produce this ideal illumination pattern, the hyperbolic subreflector in a dual-reflector Cassegrain system will yield an illumination which very closely approximates this ideal.

A radiation pattern much more typical of common feed antennas (horns, spirals, cupped dipoles, *et cetera*) is shown in Figure 8.3. Clearly, this pattern is

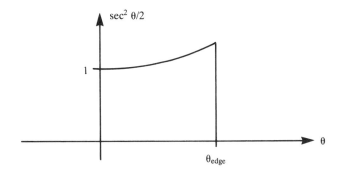

Figure 8.2 *Secant squared* feed pattern.

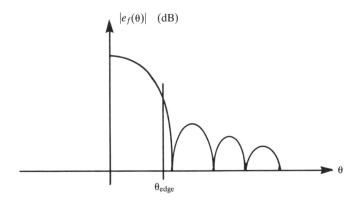

Figure 8.3 Typical feed pattern.

not at all similar to the ideal secant squared pattern. However, if pattern characteristics are chosen judiciously, this pattern type can produce high quality secondary patterns. The key is to tailor the feed pattern beamwidth to the reflector edge angle, θ_{edge}, shown in Figure 8.3. Generally, if the edge angle is chosen such that the feed pattern is 10–15 dB down from the peak valve, there will be a good tradeoff between illumination efficiency and spillover loss.

Of course, these general notions are primarily valuable for establishing an initial design. A final design would be determined by actually running an analysis program repeatedly to optimize gain, sidelobe levels, *et cetera*. We note in passing that

$$r = 2F \frac{\sin \theta}{1 + \cos \theta} \qquad (8.11)$$

gives

$$\frac{D}{4F} = \tan \frac{\theta_{edge}}{2} \qquad (8.12)$$

This relates the commonly used F/D number to the edge angle of the reflector.

8.2 LATERAL FEED DEFOCUSING AND PETZVAL SURFACE

Parabolic reflector antennas today are rarely designed to operate with only a single feed located on the axis of symmetry. Weight, space, and cost considerations will often require a single reflector to support a large number of independent antenna beams. In addition, complex "contour beam" patterns are often synthesized using multi-horn "cluster" feeds situated in the focal region of the reflector. Because of these and other considerations, a study of lateral feed defocusing effects on far-field pattern quality is central to the subject of reflector antenna design.

We begin with a ray-optics discussion of the various types of aberrations which can arise when a feed antenna is laterally and axially displaced from the focal point in a front-fed parabolic reflector antenna system [1]. Consider Figure 8.4 which shows a symmetrical parabolic reflector with displaced feed. We can see that

$$\rho' = \rho - \epsilon \qquad (8.13)$$

Let

$$\epsilon = \epsilon_x \hat{x} + \epsilon_z \hat{z} \qquad (8.14)$$

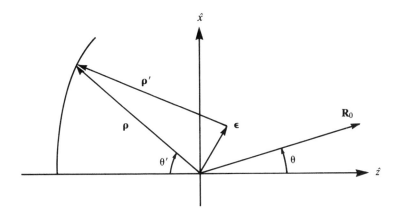

Figure 8.4 Geometry of front-fed paraboloid with defocused feed.

For a parabola,

$$\boldsymbol{\rho} = \rho(\theta') \, \hat{\rho}(\theta') \tag{8.15}$$

where

$$\rho(\theta') = \frac{2F}{1 + \cos \theta'} \tag{8.16}$$

and

$$\hat{\rho}(\theta') = \sin \theta' \cos \phi' \hat{x} + \sin \theta' \sin \phi' \hat{y} - \cos \theta' \hat{z} \tag{8.17}$$

The far-field vector, $\hat{R}_0$, is

$$\hat{R}_0 = \sin \theta \cos \phi \hat{x} + \sin \theta \sin \phi \hat{y} + \cos \theta \, \hat{z} \tag{8.18}$$

The optical path length in the direction of $\hat{R}_0$ is given by

$$\rho' - \boldsymbol{\rho}' \cdot \hat{R}_0$$

The dot product is easily calculated from the equations above. The other term is given as

$$|\boldsymbol{\rho}'| = |\boldsymbol{\rho} - \boldsymbol{\epsilon}| = \rho \left[1 + \left(\frac{\epsilon}{\rho} \right)^2 - 2 \left(\frac{\epsilon_x}{\rho} \sin \theta' \cos \phi' - \frac{\epsilon_z}{\rho} \cos \theta' \right) \right]^{1/2} \tag{8.19}$$

Using the Taylor series expansion,

$$(1 + x)^{1/2} \cong 1 + \frac{x}{2} - \frac{1}{2}\left(\frac{x}{2}\right)^2 \tag{8.20}$$

about $x = 0$, we get

$$\text{path length} = 2F + \frac{1}{2}\frac{\epsilon_x^2}{\rho} - \epsilon_x \sin \theta' \cos \phi' + \epsilon_z \cos \theta'$$

$$- \frac{1}{2}\frac{\epsilon_x^2}{\rho} \sin^2 \theta' \cos^2 \phi' - \rho \sin \theta \sin \theta' \cos (\phi - \phi') \tag{8.21}$$

$$- \rho \cos \theta' (1 - \cos \theta) + \epsilon_x \sin \theta \cos \phi + \epsilon_z \cos \theta$$

The first term and the last two terms in (8.21) are constant with respect to the aperture coordinates and may be ignored. The third term in (8.21) contains a linear phase taper as well as higher order coma aberrations. The fifth term represents the astigmatism of the system. The sixth term is the Fourier transform kernel. The remaining terms (the second, fourth and seventh) represent the phase front curvature.

One classic criterion for image quality is that the field curvature (i.e., the terms in (8.21) that are proportional to r^2, where $r = \rho \sin \theta'$) is zero. Using the following equations,

$$\rho = f[1 + (r/2F)^2] \tag{8.22}$$

$$\sin \theta' = \frac{r}{F[1 + (r/2F)^2]} \tag{8.23}$$

$$\cos \theta' = \frac{1 - (r/2F)^2}{1 + (r/2F)^2} \tag{8.24}$$

It is readily shown that the curvature is zero when

$$\epsilon_z = -\frac{1}{2F} \epsilon_x^2 \tag{8.25}$$

Equation (8.25) defines the *Petzval surface*, which is a paraboloidal surface in the focal region having a focal length one-half that of the actual reflector and bending toward the reflector. This is the locus of feed points for which (to a first order) the curvature of the far field is zero.

An interesting study on this subject is given in [2], from which the following conclusions can be drawn:

1. In general, larger F/D reflectors have maximum receive scan-gain contours which lie closer to the Petzval surface for larger amounts of lateral displacement. Smaller F/D reflectors have maximum receive scan-gain contours which depart earlier from the Petzval surface and then seem to display more of an e_x^4 type of behavior in the opposite direction (i.e., away from the reflector). This may indicate the increasing importance of phase aberrations proportional to r^4.

2. The transmit scan-gain contours also display this $\epsilon_z = -A\ \epsilon_x^2 + B\ \epsilon_x^4$ type of behavior, with higher illumination tapers yielding contours closer to the focal plane.

3. Higher scan gain is achieved when the feed is directed parallel to the axis of the reflector rather than toward the vertex, unless spillover is very high.

Figures 8.5 and 8.6 [2] indicate the maximum scan-gain contours for an F/D = .433 reflector and an F/D = .604 reflector (note that all figures from [2] incorrectly indicate the Petzval surface as bending *away* from the reflector even though it actually bends *toward* the reflector).

8.3 SYNTHESIS OF A SYMMETRICAL SINGLE REFLECTOR FOR FAR-FIELD AMPLITUDE CONTROL

In this section we consider a topic which is somewhat dissociated from the rest of the Chapter, but is worthy of note. Single reflector synthesis techniques are generally used to design unfocused reflectors which radiate energy over wide angular

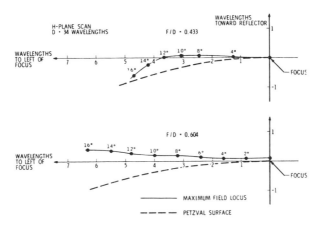

Figure 8.5 Comparison of maximum field locus and Petzval surface for receiving paraboloid (after Rusch and Ludwig, © 1973 IEEE).

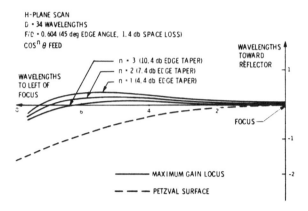

Figure 8.6 a Comparison of maximum transmit-gain contours and Petzval surface; $F/D = 0.433$ (after Rusch and Ludwig).

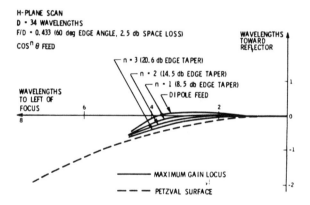

Figure 8.6 b Comparison of maximum transmit-gain contours and Petzval surface; $F/D = 0.604$ (after Rusch and Ludwig).

regions. Such an antenna might find use as a subreflector in a dual-reflector Cassegrain system or as a stand-alone antenna in an Earth coverage system such as the LANDSAT antenna shown in Figure 8.7.

Consider the symmetrical single-reflector antenna shown in Figure 8.8. It is worthwhile to note that the slope, dz/dx, of the reflector curve determines θ_2 as a function of θ_1 (i.e., it determines the reflected field direction). Similarly, the curvature, dz^2/dx^2, of the curve determines the radius of curvature of the reflected wavefront (i.e., the reflected field amplitude). We seek to determine the position, slope, and curvature of the reflector surface which will produce the desired radiation pattern.

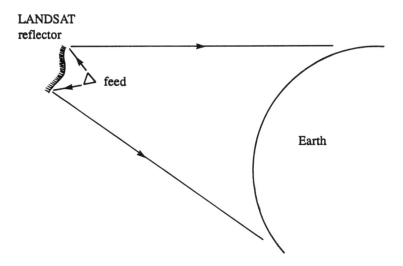

Figure 8.7 LANDSAT Earth-coverage reflector.

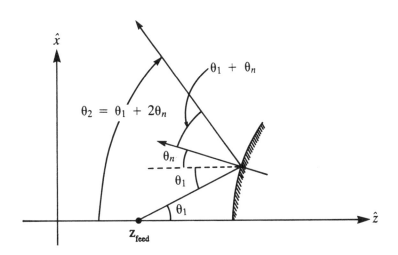

Figure 8.8 Symmetrical single-reflector antenna.

Suppose that the feed power pattern is of the form

$$E^2(\theta_1)$$

and the desired radiation power pattern is of the form

$$A^2 u^2(\theta_2)$$

where

$$|u^2(\theta_2)|_{max} = 1 \qquad (8.26)$$

The constant, A, is determined from the condition

$$A^2 = \frac{\displaystyle\int_0^{\theta_{1,max}} E^2(\theta_1) \sin \theta_1 \, d\theta_1}{\displaystyle\int_0^{\theta_{2,max}} u^2(\theta_2) \sin \theta_2 \, d\theta_2} \qquad (8.27)$$

where $E(\theta_1)$ is known and $u(\theta_2)$ is prescribed.

Expressing this power law in differential form gives

$$\frac{d\theta_2}{d\theta_1} = \frac{E^2(\theta_1) \sin \theta_1}{A^2 u^2(\theta_2) \sin \theta_2} \qquad (8.28)$$

Since $\theta_2 = \theta_1 + 2\theta_n$, (8.28) becomes

$$\frac{d\theta_2}{d\theta_1} = \frac{E^2(\theta_1) \sin \theta_1}{A^2 u^2(\theta_1 + 2\theta_n) \sin (\theta_1 + 2\theta_n)} \qquad (8.29)$$

where

$$\theta_n = \tan^{-1} dz/dx \qquad (8.30)$$

Equation (8.29) is an equation for $d\theta_2/d\theta_1$ derived strictly from conservation of power considerations. We now derive a second equation for $d\theta_2/d\theta_1$ from geometrical considerations. Since

$$\theta_2 = \theta_1 + 2\theta_n \qquad (8.31)$$

we have

$$\frac{d\theta_2}{d\theta_1} = 1 + 2 \frac{d\theta_n}{d\theta_1} \qquad (8.32)$$

By (8.30), we have

$$\frac{d\theta_2}{d\theta_1} = 1 + 2 \frac{d}{dx}\left[\tan^{-1} \frac{dz}{dx}\right] \cdot \frac{dx}{d\theta_1} \qquad (8.33)$$

However, since

$$\theta_1 = \tan^{-1}\left[\frac{x}{z - z_f}\right] \tag{8.34}$$

we get

$$\frac{d\theta_1}{dx} = \frac{\left(z - z_f - x\frac{dz}{dx}\right)}{(z - z_f)^2\left[1 + \left(\frac{x}{z - z_f}\right)^2\right]} \tag{8.35}$$

or

$$\frac{dx}{d\theta_1} = \frac{(z - z_f)^2\left[1 + \left(\frac{x}{z - z_f}\right)^2\right]}{\left(z - z_f - x\frac{dz}{dx}\right)} \tag{8.36}$$

with this, (8.33) becomes

$$\frac{d\theta_2}{d\theta_1} = 1 + 2\left[1 + \left(\frac{dz}{dx}\right)^2\right]^{-1} \cdot \frac{d^2z}{dx^2} \cdot \frac{(z - z_f)^2\left[1 + \left(\frac{x}{z - z_f}\right)^2\right]}{(z - z_f - x\,dz/dx)} \tag{8.37}$$

Now equate (8.29) and (8.37) to get

$$\frac{E^2\left(\tan^{-1}\frac{x}{z - z_f}\right)\sin\left(\tan^{-1}\frac{x}{z - z_f}\right)}{A^2u^2\left(\tan^{-1}\frac{x}{z - z_f} + 2\tan^{-1}\frac{dz}{dx}\right)\sin\left(\tan^{-1}\frac{x}{z - z_f} + 2\tan^{-1}\frac{dz}{dx}\right)}$$

$$= 1 + 2\frac{dz^2}{dx^2}\frac{(z - z_f)^2\left[1 + \left(\frac{x}{z - z_f}\right)^2\right]}{[1 + (dz/dx)^2]\left(z - z_f - x\frac{dz}{dx}\right)} \tag{8.38}$$

Equation (8.38) is of the form

$$\frac{d^2z}{dx^2} = f[x, z(x), z'(x)] \tag{8.39}$$

In other words it is a second-order ordinary differential equation for $z(x)$. In order to solve it, the initial values for $z(x)$ and $z'(x)$ must be known at $x = R$, the radius of the reflector. These are easily obtained because $z(R)$ is assumed known and

$$z'(R) = \tan \frac{1}{2}(\theta_{2,\text{edge}} - \theta_{1,\text{edge}}) \tag{8.40}$$

Note that this has been strictly a scalar analysis and does not include polarization effects.

REFERENCES

1. Ruze, J., "Lateral Feed Displacement in a Paraboloid," *IEEE Trans. Antennas Propagat.*, Vol. AP-13, pp. 660–665, September 1965.
2. Rusch, W.V.T. and A.C. Ludwig, "Determination of the Maximum Scan-Gain Contours of a Beam-Scanning Paraboloid and Their Relation to the Petzval Surface," *IEEE Trans. Antennas Propagat.*, Vol. AP-21, pp. 141–147, March 1973.

Chapter 9
Design of Dual-Reflector Antenna Systems

In the present context, *dual-reflector* will refer either to the classical Cassegrain system shown in Figure 9.1 or some shaped or offset version of this basic scheme. Since the field of dual-reflector design can be quite specialized, this chapter will primarily consider those aspects which would be of general interest. The more detailed aspects are treated in the References.

9.1 GEOMETRICAL RELATIONSHIPS FOR CASSEGRAIN REFLECTORS

Consider Figure 9.1 which schematically illustrates a symmetrical Cassegrain system. Only four numbers are necessary to specify the entire Cassegrain system. Two parameters are necessary to specify the main parabolic reflector. These may be any two of the three parameters F, R_2, θ_2. Ordinarily, three of the five numbers e, c, R_1, θ_1, θ_2 would be required to specify the hyperbolic subreflector. In this case however, since the hyperbola must be compatible with both the parabolic main reflector and the feed dimensions, two of these three numbers will be determined by the main reflector and the third will be determined by the lateral extent of the feed.

Referring to Figure 9.2, the design process might proceed as follows. First of all, this system naturally will be designed to operate at some particular frequency. Generally, the primary constraint imposed on an antenna system is the on-axis gain. Knowing the operating frequency and the specified on-axis gain, the main reflector diameter now may be determined using (1.26). In addition, the feed parameters will undoubtedly be known at the given operating frequency. Hence, θ and R_f will be known. Lastly, the main reflector focal length, F, will also likely be known, either from dimensional constraints or from scan performance considerations. So, the four parameters R_1, θ, R_f, F will likely be the starting point for a Cassegrain design.

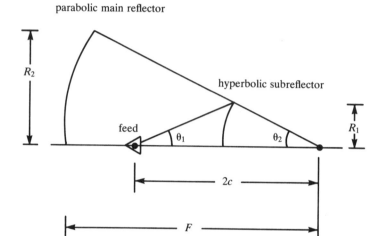

Figure 9.1 Classic symmetrical Cassegrain system.

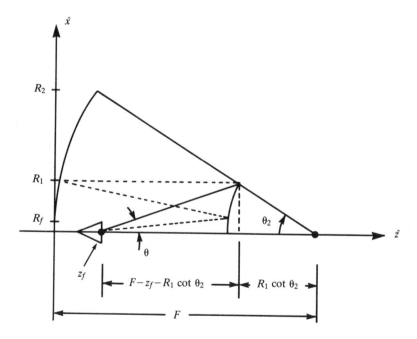

Figure 9.2 On the Cassegrain design process.

Next it is necessary to position the feed along the z-axis. A common criterion is to locate z_f so that the dotted line in Figure 9.2 just passes through the widest point of the feed. In this case, the feed blockage will equal the subreflector blockage. This means that

$$\frac{R_1}{F - \dfrac{R_1^2}{4F}} = \frac{R_f}{F - z_f} \tag{9.1}$$

A good approximation to (9.1) is

$$R_1 = R_f\left(\frac{F}{F - z_f}\right) \tag{9.2}$$

A second equation can also be written relating (R_1, z_f):

$$R_1 = (F - z_f - R_1 \cot \theta_2) \tan \theta \tag{9.3}$$

or,

$$R_1 = \frac{F - z_f}{\cot \theta + \cot \theta_2} \tag{9.4}$$

Equations (9.2) and (9.4) are two equations in the two unknowns (R_1, z_f). Equating (9.2) and (9.4) yields

$$(F - z_f)^2 = F \cdot R_f(\cot \theta + \cot \theta_2) \tag{9.5}$$

which can be solved for z_f. Then R_1 can be solved for using (9.2) or (9.4).

In some cases, it may not be practical to impose the criterion that feed blockage will equal the subreflector blockage. In that case, (9.4) can still be used to produce a family of $(z_f\ R)$, pairs, from which a practical design can be extracted. Now consider Figure 9.3. The eccentricity of the hyperbola may be calculated from the equation

$$\rho_1 - \rho_2 = \frac{2c}{e} \tag{9.6}$$

which defines the hyperbola. This yields

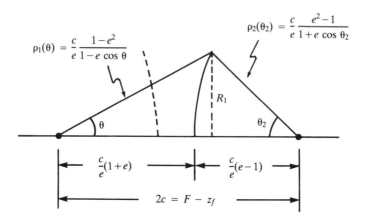

Figure 9.3 Hyperbolic subreflector geometry.

$$e = \frac{2c}{\rho_1 - \rho_2} \tag{9.7}$$

or

$$e = \frac{F - z_f}{R_1 (\csc \theta - \csc \theta_2)} \tag{9.8}$$

and the hyperbola (as well as the parabola) is completely specified.

The design of an offset Cassegrain system proceeds along essentially the same lines as for a symmetrical system. In fact, many of the design equations for the symmetrical configuration are directly applicable to the offset configuration. Some new equations, however, will be presented which are specific to the offset case. Most of these have rather involved derivations which we will not go into here. Further discussion of these design equations may be found in the literature [1–3].

The primary advantage of offset Cassegrain systems is the absence of feed blockage, while the primary disadvantage in the past has been high levels of cross-polarization due to the asymmetry of the structure. However, the conventional offset configuration may be modified [4] so that on-axis cross-polarization is eliminated. This modified offset Cassegrain design, described in a later section, may also be designed using the equations herein for conventional offset systems.

Looking at Figure 9.4, the following observations may be made. First, the offset paraboloid can be obtained either a) by intersecting the paraboloid of revolution with a circular cylinder of radius $D/2$ which is displaced by a distance, d, from the z-axis, or b) by intersecting the paraboloid of revolution with a circular

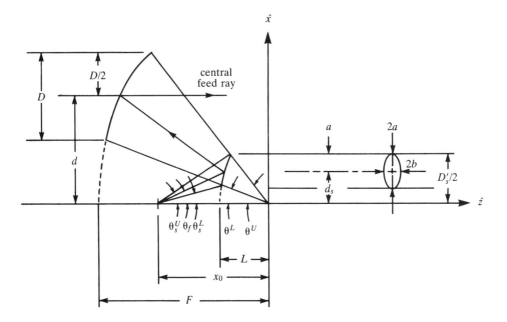

Figure 9.4 Geometrical relationships for offset Cassegrain antenna.

cone of half-angle $(\theta^U - \theta^L)/2$, which is tilted by an angle of $(\theta^U + \theta^L)/2$ with respect to the z-axis. The two descriptions are equivalent.

The various main reflector parameters are related by the following equations [2]:

$$\theta^B = \frac{1}{2} (\theta^U + \theta^L) \tag{9.9}$$

$$\theta^C = \frac{1}{2} (\theta^U - \theta^L) \tag{9.10}$$

$$D = \frac{4F \sin \theta^C}{\cos \theta^B + \cos \theta^C} \tag{9.11}$$

$$d = \frac{2F \sin \theta^B}{\cos \theta^B + \cos \theta^C} \tag{9.12}$$

$$\theta^{U,L} = 2 \tan^{-1} \frac{d \pm D/2}{2F} \tag{9.13}$$

The edge angles of the hyperbolic subreflector are calculated from the main reflector parameters as before in the symmetrical case. The offset subreflector is obtained either by a) intersecting the hyperboloid of revolution by a circular cone of half-angle $(\theta_S^U + \theta_S^L)/2$ with respect to the z-axis, or b) intersecting the hyperboloid of revolution with an elliptical cylinder of radii a, b which is displaced a distance d_S from the z-axis. Both of these descriptions are equivalent.

The various subreflector parameters are related by the following equations [2]:

$$M = \frac{\tan \theta^U/2}{\tan \theta_S^U/2} = \frac{\tan \theta^L/2}{\tan \theta_S^L/2} = \frac{\tan \theta/2}{\tan \theta_S/2} \tag{9.14}$$

$$e = \frac{M + 1}{M - 1} \tag{9.15}$$

$$\theta_S^B = \frac{1}{2}(\theta_S^U + \theta_S^L) \tag{9.16}$$

$$\theta_S^C = \frac{1}{2}(\theta_S^U - \theta_S^L) \tag{9.17}$$

$$\frac{z_0}{L} = 1 + M \tag{9.18}$$

$$d_S = \frac{(1 + e)(e \cos \theta^B + \cos \theta^C) \sin \theta^B}{(1 - e^2) \sin^2 \theta^B + (\cos \theta^B + e \cos \theta^C)^2} \tag{9.19}$$

$$\frac{z_0}{D_S'/2} = \cot \theta^U + \cot \theta_S^U \tag{9.20}$$

$$\frac{z_0}{D_S'/2 - 2a} = \cot \theta^L + \cot \theta_S^L \tag{9.21}$$

$$\frac{a^2}{L^2} = \frac{(1 + e)^2(\cos \theta^B + e \cos \theta^C)^2 \sin^2 \theta^C}{[(1 - e^2) \sin^2 \theta^B + (\cos \theta^B + e \cos \theta^C)^2]^2} \tag{9.22}$$

$$\frac{b^2}{L^2} = \frac{(1 + e)^2 \sin^2 \theta^C}{(1 - e^2) \sin^2 \theta^B + (\cos \theta^B + e \cos \theta^C)^2} \tag{9.23}$$

The feed tilt angle, θ_f, will most likely not be equal to $(\theta_S^U + \theta_S^U)/2$, the tilt angle of the subreflector intersecting cone. The angle θ_f is chosen so that the central feed ray at θ_f intersects the center of the projection of the main reflector aperture onto the x-y plane. This is shown in Figure 9.4.

The intersection of a circular cone with either the paraboloid or the hyperboloid of revolution produces a curve which defines the rim of the reflector. This curve is planar [1] in both cases. First consider the offset paraboloid shown in Figure 9.5. It is an easy matter to show that the rim is a planar curve. The equation of the parabola is

$$z = \frac{1}{4F}(x^2 + y^2) - F \tag{9.24}$$

and the equation of the cone is

$$(x^2 + y^2 + z^2) \cos^2 \alpha = (z \cos \beta - x \sin \beta)^2 \tag{9.25}$$

By (9.24),

$$z + \frac{z^2}{4F} = \frac{1}{4F}(x^2 + y^2 + z^2) - F \tag{9.26}$$

Substituting (9.25) into (9.26) gives

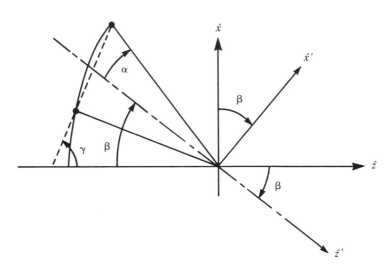

Figure 9.5 Intersection of circular cone and paraboloid of revolution.

$$x = \left[\frac{\cos \alpha + \cos \beta}{\sin \beta}\right] z + 2F \frac{\cos \alpha}{\sin \beta} \tag{9.27}$$

from which we see that the angle, γ, in Figure 9.5 is given by

$$\tan \gamma = \frac{\cos \alpha + \cos \beta}{\sin \beta} \tag{9.28}$$

The intersection of a titled circular cone with a hyperboloid of revolution also produces a planar curve [3]. Consider Figure 9.6, which illustrates the problem schematically. In this case, the plane of the rim is tilted with respect to the x-y plane by an angle, θ_E, given by

$$\tan \theta_E = \frac{\cos \theta_T'' - e \cos \theta_A''}{\sin \theta_T''} = \frac{\cos \theta_T' + e \cos \theta_A'}{\sin \theta_T'} \tag{9.29}$$

where all angles are as shown in Figure 9.6.

The rim is a planar ellipse in the θ_E plane. The eccentricity, e_1, of this ellipse is

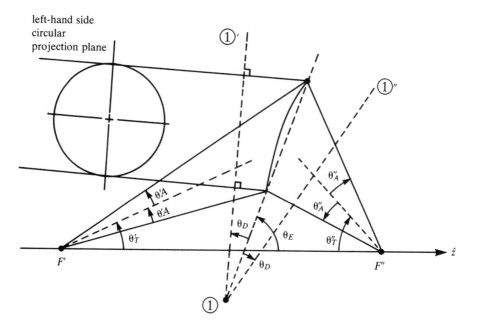

Figure 9.6 Geometry of an offset hyperbolic subreflector.

$$e_1 = e \cos \theta_E \tag{9.30}$$

where e is the eccentricity of the hyperboloid of revolution.

The two planes $\textcircled{1} - \textcircled{1}'$ and $\textcircled{1} - \textcircled{1}''$ are the *circular projection planes,* i.e., the planes in which the projection of the elliptical rim is a circle. They are tilted by an angle, θ_D, with respect to the plane of the rim where θ_D is given by the equation

$$\cos \theta_D = \sqrt{1 - e_1^2} \tag{9.31}$$

9.2 SYNTHESIS OF SYMMETRICAL DUAL-SHAPED REFLECTOR ANTENNA SYSTEMS

We seek to design a Cassegrain-like reflector system having a uniform phase distribution and a prescribed amplitude distribution across the exit aperture of the main reflector. To this end, consider Figure 9.7a which illustrates the dual reflector system schematically.

Our goal is to obtain equations for the subreflector and main reflector in this dual-shaped system. The approach used is a geometrical optics ray-tracing process. It is assumed initially that the radii R_1, R_2 of the two reflectors are known, as are the edge angles from feed-to-subreflector and sub-to-main reflector.

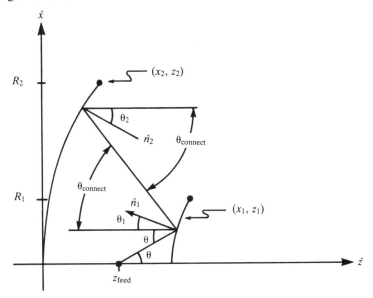

Figure 9.7a Symmetrical dual-shaped antenna system.

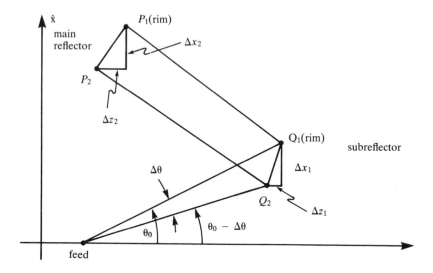

Figure 9.7b On the algorithm for dual-reflector shaping.

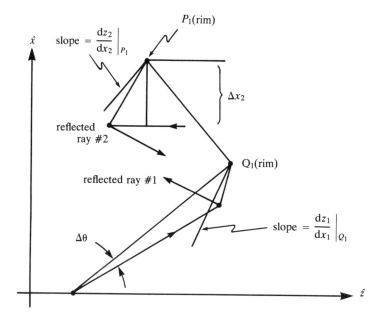

Figure 9.7c Alternate procedure for dual-reflector shaping.

ray #2   ray #1

Figure 9.7d Criterion for search procedure.

Before actually beginning the analysis, it is worthwhile to note that the uniform aperture phase condition is actually implied in the way Figure 9.7a is drawn. That is, if all exit rays are parallel to the symmetry axis, then they are all automatically in-phase since the phase fronts remain perpendicular to the rays after any number of reflections. This knowledge will simplify the analysis immensely because it eliminates the need to trace ray trajectories through the dual-reflector system and then equate their length to some constant valve.

The aperture phase constraint now eliminated we can examine the aperture amplitude constraint. For the purposes of geometrical optics, power is regarded as flowing within *ray tubes;* therefore total power is conserved within a ray tube. Hence,

$$E^2(\theta) \sin \theta \; d\theta \; d\phi = A^2(x_2)x_2 \; dx_2 \; d\phi_2 \tag{9.32}$$

where $A(x_2)$ is the prescribed aperture amplitude distribution and $E(\theta)$ is the (assumed symmetrical) feed radiation pattern. By symmetry, $d\phi = d\phi_2$, hence (9.32) becomes

$$E^2(\theta) \sin \theta \; d\theta = A^2(x_2)x_2 \; dx_2 \tag{9.33}$$

We can say that

$$A(x_2) = Au(x_2) \tag{9.34}$$

where

$$|u(x_2)|_{max} = 1 \tag{9.35}$$

Then, (9.33) becomes

$$E^2(\theta) \sin \theta \; d\theta = A^2 u^2(x_2)x_2 \; dx_2 \tag{9.36}$$

As in Section 8.3, this gives

$$A^2 = \frac{\displaystyle\int_0^{\theta_{edge}} E^2(\theta) \sin \theta \; d\theta}{\displaystyle\int_0^{R_2} u^2(x_2)x_2 \; dx_2} \tag{9.37}$$

where θ_{edge} = subreflector edge angle. Equation (9.36) can be rewritten as

$$\frac{dx_2}{d\theta} = \frac{E^2(\theta)\,\sin\,\theta}{A^2 u^2(x_2) x_2} \tag{9.38}$$

This is an expression for $dx_2/d\theta$ in terms of known and prescribed quantities. We now derive a second expression for $dx_2/d\theta$ in terms of the geometrical properties of the two reflector curves. To begin, we note that

$$\frac{dx_2}{d\theta} = \frac{dx_2}{d\theta_{connect}} \times \frac{d\theta_{connect}}{d\theta} \tag{9.39}$$

Equating (9.38) and (9.39) gives

$$\frac{E^2(\theta)\,\sin\,\theta}{A^2 u^2(x_2) x_2} = \frac{dx_2}{d\theta_{connect}} \cdot \frac{d\theta_{connect}}{d\theta} \tag{9.40}$$

or

$$E^2(\theta)\,\sin\,\theta\,\frac{d\theta}{d\theta_{connect}} = A^2 u^2(x_2) x_2 \frac{dx_2}{d\theta_{connect}} \tag{9.41}$$

This last equation could have actually been derived directly from (9.33) by merely dividing both sides by $d\theta_{connect}$. Hence we see that (9.41) can be written as

$$\frac{d\ (\text{feed power})}{d\theta_{connect}} = \frac{d\ (\text{aperture power})}{d\theta_{connect}} \tag{9.42}$$

It can easily be shown that

$$\frac{d\theta_{connect}}{d\theta} = 1 + 2\frac{d^2 z_1}{dx_1^2}\frac{\left[1 + \left(\dfrac{x_1}{z_1 - z_f}\right)^2\right](z_1 - z_f)^2}{[1 + (dz_1/dx_1)^2]\left(z_1 - z_f - x_1\dfrac{dz_1}{dx_1}\right)} \tag{9.43}$$

and

$$\frac{dx_2}{d\theta_{connect}} = \frac{1 + \left(\dfrac{dz_2}{dx_2}\right)^2}{2\dfrac{d^2 z_2}{dx_2^2}} \tag{9.44}$$

Hence,

$$\frac{d\theta}{d\theta_{connect}} = f(x_1, z_1, z_1', z_1'') \tag{9.45}$$

$$\frac{dx_2}{d\theta_{connect}} = g(x_2, z_2', z_2'') \tag{9.46}$$

and (9.41) can be separated to yield

$$E^2(\theta) \sin \theta \, f(x_1, z_1, z_1', z_1'') = \frac{d \, (power)}{d\theta_{connect}} \tag{9.47}$$

and

$$A^2 u^2(x_2) x_2 g(x_2, z_2'') = \frac{d \, (power)}{d\theta_{connect}} \tag{9.48}$$

where

$$\theta = \tan^{-1}\left(\frac{x_1}{z_1 - z_f}\right) \tag{9.49}$$

Equations (9.47) and (9.48) are the two ordinary differential equations for the sub and main reflectors. The right-hand sides of each are equal and subject to the constraint that

$$\int_0^{\theta_{connect,max}} [d \, (power)/d\theta_{connect}] \, d\theta_{connect} = \int_0^{\theta_{edge}} E^2(\theta) \sin \theta \, d\theta \tag{9.50}$$

The two ordinary differential equations are second order and may be solved by assuming initial values of

$$z(x = R) \text{ and } dz/dx|_{x=R}$$

which correspond to an unshaped Cassegrain system. The algorithm for solving this set of ordinary differential equations can be stated with the aid of Figure 9.7b. The independent variable can be x_2. So,

1. Choose Δx_2. Now calculate $\Delta\theta$ according to (9.33), the conservation of power equation.
2. Since $dz_2/dx_2|P_1$ is known from the initial conditions at the rim of the main reflector, Δz_2 can be calculated once Δx_2 is chosen, locating the point P_2.

3. In addition, $dz_1/dx_1|Q_1$ is also known from the initial conditions at the rim of the subreflector. We then can locate the point Q_2 as the intersection of the line from the feed at angle $\theta_0 - \Delta\theta$, and the line from Q_1 having the slope, $dz_1/dx_1|Q_1$. Now we know Δz_1.
4. Knowing the new points P_2, Q_2, we can calculate $\theta_{connect}$, the angle of the line connecting P_2, Q_2, by the equation

$$\tan \theta_{connect} = \frac{x_2 - x_1}{z_1 - z_2}$$

5. Now the quantity $\Delta(\text{power})/\Delta\,\theta_{connect}$ at P_1, Q_1 can be calculated (this is the forcing function for (9.47) and (9.48)).
6. This allows z_1'', z_2'' to be calculated (at P_1, Q_1).
7. Now, $z_1'\,|Q_2$ and $z_2'|P_2$ can be calculated.
8. Now we repeat the iteration process all over again.

One additional method can also be employed to determine the symmetrical dual-shaped reflector system. This method is very simple and can be understood with the help of Figure 9.7c. We can choose Δx_2 and then calculate $\Delta\theta$ from conservation of power. We know the values of

$$\frac{dz_2}{dx_2}\bigg|_{P_1} \text{ and } \frac{dz_1}{dx_1}\bigg|_{Q_1}$$

from the initial conditions at the rim of each reflector. However, we do not know the second derivatives,

$$\frac{d^2z_z}{dx_2^2}\bigg|_{P_1} \text{ and } \frac{d^2z_1}{dx_1^2}\bigg|_{Q_1}$$

at each rim. We can determine these quantities, however, using a search procedure and assuming initial values for the search to be the second derivatives for an unshaped parabolic/hyperbolic Cassegrain system. The goal of the search is to make rays 1 and 2 co-linear as shown in Figure 9.7d. Note that the two rays must not only be parallel, but also coincident.

One last comment is in order, with respect to dual-reflector systems. Even though the reflector designed by geometrical optics methods has the subreflector edge ray and main reflector edge ray being coincident, in practice improved performance may be achieved by enlarging the subreflector by a wavelength or so beyond this design limit. This is referred to as "oversizing" the subreflector. This is useful because diffraction effects at the subreflector edge cause the reflected fields to decay in the vicinity of the main reflector rim. Slightly oversizing the subreflector will ensure that the main reflector is efficiently illuminated near the rim. Note that too much oversizing may result in increased subreflector blockage and perhaps also increase spillover loss at the main reflector rim. Again, a reflector analysis computer program will help in determining an exact design.

9.3 OFFSET DUAL-REFLECTOR SYSTEMS HAVING ZERO CROSS-POLARIZATION

Offset dual-reflector systems are very attractive to antenna designers because they offer the prospect of building an antenna system having no feed blockage. On the other hand, the asymmetry of the offset configuration may result in high levels of cross-polarized radiation unless steps are taken to mitigate this effect. In this section, we present one effective solution to this problem [4]. Consider the offset dual-reflector systems shown in Figures 9.8 and 9.9.

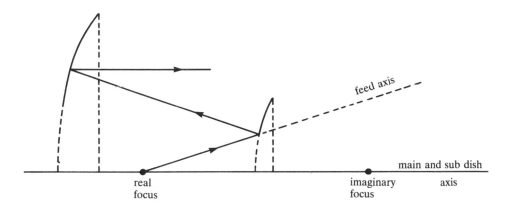

Figure 9.8 Conventional offset Cassegrain configuration.

In any offset Cassegrain system where low levels of cross-polarization are to be obtained, the main reflector symmetry axis has to be tilted with respect to the symmetry axis of the original symmetrical parent Cassegrain system. This modified offset Cassegrain configuration (shown in Figure 9.9) is to be compared with the conventional offset Cassegrain configuration shown in Figure 9.8. In this section, we derive the equation for the main reflector tilt angle as a function of the feed tilt angle.

The derivation of this equation is based on a principle put forth by Dragone [5] which states that in order for the reflector system to be rotationally symmetrical,

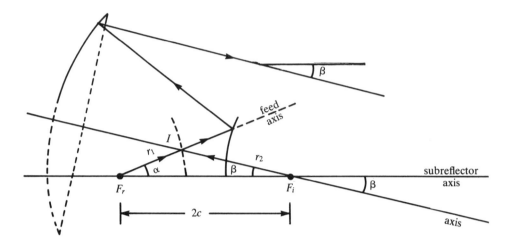

Figure 9.9 Modified offset Cassegrain system.

the direction of the central feed ray must be unchanged after successive reflections from the subreflector, the main reflector, infinity, and the complementary subreflector surface. Clearly, this condition will be met when the angles (α, β) are chosen so that the ray at angle α from the real focus, F_r, meets the ray at angle β from the imaginary focus, F_i, at an intersection point, I, on the unused complementary subreflector sheet, as shown in Figure 9.9.

Applying the law of sines to triangle $F_r F_i I$ gives

$$\frac{\sin \alpha}{r_2} = \frac{\sin (\alpha + \beta)}{2c} = \frac{\sin \beta}{r_1} \tag{9.51}$$

where

$$r_1 = \frac{c}{e} \frac{e^2 - 1}{1 + e \cos \alpha} \tag{9.52}$$

$$r_2 = \frac{c}{e} \frac{1 - e^2}{1 - e \cos \beta} \tag{9.53}$$

and $e > 1$ for the hyperboloid. The first equality in (9.51) gives

$$\tan \alpha = \frac{(e^2 - 1) \sin \beta}{(e^2 + 1) \cos \beta - 2e} \tag{9.54}$$

and the second gives

$$\tan \beta = \frac{(e^2 - 1) \sin \alpha}{(e^2 + 1) \cos \alpha + 2e} \tag{9.55}$$

REFERENCES

1. Jamnejad-Dailami, V. and Y. Rahmat-Samii, "Some Important Geometrical Features of Conic Section Generated Offset Reflector Antennas," *IEEE Trans. Antennas Propogat.*, Vol. AP-28, pp. 952–957, November 1980.
2. Rahmat-Samii, Y., "Subreflector Extension for Improved Efficiencies in Cassegrain Antennas— GTD/PO Analysis," *IEEE Trans. Antennas Propagat.*, Vol. AP-34, pp. 1266–1269, October 1986.
3. Pogorzelski, R.J., "On the Numerical Analysis of the Subreflector of an Offset Cassegrain Microwave Antenna," *IEEE Trans. Antennas Propagat.*, Vol. AP-32, pp. 595–601, June 1984.
4. Shore, R.A., "A Simple Derivation of the Basic Design Equation for Offset Dual Reflector Antennas with Rotational Symmetry and Zero Cross Polarization," *IEEE Trans. Antennas Propagat.*, Vol. AP-33, pp. 114–116, January 1985.
5. Dragone, C., "Offset Multireflector Antennas with Perfect Pattern Symmetry and Polarization Discrimination," Bell Syst. Tech. J, Vol. 57, pp. 2663–2684, September 1978.

Suggestions for Additional Reading

1. Galindo-Israel, V., "Design of Dual-Reflector Antennas with Arbitrary Phase and Amplitude Distributions," *IEEE Trans. Antennas Propagat.*, Vol. AP-12, pp. 403–408, July 1964.
2. Galindo-Israel, V., R. Mittra, and A.G. Cha, "Aperture Amplitude and Phase Control of Offset Dual Reflectors, "*IEEE Trans. Antennas Propagat.*, Vol. AP-27, pp. 154–164, March 1979.
3. Lee, J.J., L.I. Parad, and R.S. Chu, "A Shaped Offset-Fed Dual-Reflector Antenna," *IEEE Trans. Antennas Propagat.*, Vol. AP-27, pp. 165–171, March 1979.
4. Takano, T., E. Ogawa, S. Betsudan, and S. Sato, "High Efficiency and Low Sidelobe Design for a Large Aperture Offset Reflector Antenna," *IEEE Trans. Antennas Propagat.*, Vol. AP-28, pp. 460–466, July 1980.
5. Narasimhan, M.S., V. Anantharam, and K.M. Prasad, "A Note on the Shaping of Dual Reflector Antennas," *IEEE Trans. Antennas Propagat.*, Vol. AP-29, pp. 551–552, May 1981.
6. Mittra, R., F. Hyjazie, and V. Galindo-Israel, "Synthesis of Offset Dual Reflector Antennas Transforming a Given Feed Illumination Pattern into a Specified Aperture Distribution," *IEEE Trans. Antennas Propagat.*, Vol. AP-30, pp. 251–259, March 1982.
7. Galindo-Israel, V. and R. Mittra, "Synthesis of Offset Dual Shaped Subreflector Antennas for Control of Cassegrain Aperture Distributions," *IEEE Trans. Antennas Propagat.*, Vol. AP-32, pp. 86–92, January 1984.
8. Galindo-Israel, V., W.A. Imbriale, and R. Mittra, "On the Theory of the Synthesis of Single and Dual Offset Shaped Reflector Antennas," *IEEE Trans. Antennas Propagat.*, Vol. AP-35, pp. 887–896, August 1987.

Chapter 10
Aperture Field Integration Methods

Implicit in the *geometrical optics* (GO) design approach for Cassegrain systems is the assumption that the far-field radiation pattern is directly related to (i.e., the Fourier transform of) the GO electric field over the exit aperture of the reflector. In this chapter, we return one final time to the subject of antenna analysis and investigate the validity of this assumption. In addition, we will complete this book by integrating a number of analysis and design concepts.

10.1 ELECTROMAGNETICS BACKGROUND

Consider Figure 10.1 which shows an infinite 3-dimensional volume separated by an infinite plane at $z = 0$ over which the tangential electric field is known. In the source-free region $z > 0$, the vector functions **E, H, A** all satisfy the homogeneous Helmholtz equation:

$$\nabla^2 \begin{Bmatrix} \mathbf{E} \\ \mathbf{H} \\ \mathbf{A} \end{Bmatrix} + k^2 \begin{Bmatrix} \mathbf{E} \\ \mathbf{H} \\ \mathbf{A} \end{Bmatrix} = 0 \tag{10.1}$$

It is well known that in rectangular coordinates, this equation can be separated so that a general solution is represented as a summation of elementary product solutions of the form

$$\mathbf{E}(x, y, z) = \mathbf{E}\,(\alpha, \beta)\Big|_{z=0} e^{-j(\alpha x + \beta y + \gamma z)} \tag{10.2}$$

where

$$\gamma = \pm \sqrt{k^2 - \alpha^2 - \beta^2} \tag{10.3}$$

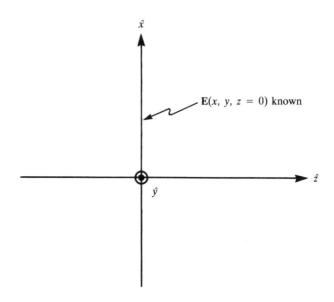

Figure 10.1 Infinite planar boundary over which the tangential electric field is known.

by the separability condition. In (10.3) we select the root having a non-negative real part and non-positive imaginary part. The plus sign is for $+z$ traveling waves and the minus sign is for $-z$ traveling waves.

The general solution for the electric field is

$$\mathbf{E}(x, y, z) = \int\!\!\int_{-\infty}^{\infty} \mathbf{E}(\alpha, \beta)\Big|_{z=0} e^{-j(\alpha x + \beta y + \gamma z)} \, d\alpha \, d\beta \tag{10.4}$$

where

$$\mathbf{E}(\alpha, \beta)\Big|_{z=0} = \text{Fourier transform of } \mathbf{E}(x, y)$$

Equation (10.4) expresses the electric field as a spectrum of plane waves:

$$\mathbf{E}(\alpha, \beta) = \frac{1}{(2\pi)^2} \int\!\!\int_{-\infty}^{\infty} \mathbf{E}(x, y) \, e^{j(\alpha x + \beta y)} \, dx \, dy \tag{10.5}$$

Equation (10.5) is deduced almost immediately from (10.4). If (10.4) is specialized to $z = 0$ and both sides are multiplied by $e^{j(\alpha' x + \beta' y)}$ and integrated over the $z = 0$ plane, we get

$$\iint_{-\infty}^{\infty} \mathbf{E}(x, y) \, e^{j(\alpha'x + \beta'y)} \, dx \, dy$$

$$= \iint_{-\infty}^{\infty} \mathbf{E}(\alpha, \beta) \, d\alpha \, d\beta \int_{-\infty}^{\infty} e^{j(\alpha'-\alpha)x} \, dx \int_{-\infty}^{\infty} e^{j(\beta'-\beta)y} \, dy$$

$$= (2\pi)^2 \iint_{-\infty}^{\infty} \mathbf{E}(\alpha, \beta) \;\; \delta(\alpha - \alpha') \, \delta(\beta - \beta') \, d\alpha \, d\beta \tag{10.6}$$

$$= (2\pi)^2 \, \mathbf{E}(\alpha', \beta')$$

Now we will use (10.4) to calculate the electric field at a point (x, y, z) far removed from the $z = 0$ plane, in order to prove that it is proportional to $\mathbf{E}$ (α, β). Equation (10.4) can be rewritten as

$$\mathbf{E}(\mathbf{r}) = \iint_{-\infty}^{\infty} \mathbf{E}(\alpha, \beta) \, e^{-j\mathbf{k}\cdot\mathbf{r}} \, d\alpha \, d\beta \tag{10.7}$$

where

$$\mathbf{k} = \alpha\hat{x} + \beta\hat{y} + \gamma\hat{z} \tag{10.8}$$

and

$$\mathbf{r} = \frac{r}{k} (\alpha_0\hat{x} + \beta_0\hat{y} + \gamma_0\hat{z}) \tag{10.9}$$

where, in 10.9,

$$r = \text{radial distance to far-field point}$$
$$\alpha_0/k, \; \beta_0/k, \; \gamma_0/k = \text{direction cosines of far-field point}$$

The integral in (10.7) can be evaluated in closed form as $r \to \infty$. To perform the integration, we employ a technique known as the *stationary phase method*. This is an asymptotic integration technique which (in this case) yields an integral that becomes exact as $r \to \infty$.

The idea behind the stationary phase method is that in the region where $\mathbf{k}$ and $\mathbf{r}$ are nearly parallel, the exponent in (10.7) varies slowly (i.e., it's nearly constant). In fact, since the dot product varies as the cosine of the angle between $\mathbf{k}$ and $\mathbf{r}$, the phase will have a local minimum in the direction where $\mathbf{k} \parallel \mathbf{r}$. Away from this point, $\mathbf{k} \cdot \mathbf{r}$ will vary rapidly and the complex phase term will oscillate rapidly. This causes cancellation in the integral so that the region in k-space away from $\mathbf{r}$ will contribute little to the final integral. At a given frequency (constant value of k), this region will become more and more localized in the vicinity of the observation vector, $\mathbf{r}$, as $r \to \infty$.

In physical terms as $r \to \infty$, only the plane wave in the direction of $\mathbf{r}$ (out of the entire spectrum of plane waves) will contribute to the far-zone electric field at the observation point, $\mathbf{r}$. Intuitively, we assume that the far-zone field strength is proportional to the radiation pattern in the direction of the field point. In this case however it is necessary to prove the obvious since we are now dealing with electric field sources rather than electric current sources. (Had we employed the concept of magnetic current sources, we could have bypassed this entire stationary phase process.)

The phase term in (10.7) can be written as

$$\mathbf{k} \cdot \mathbf{r} = \frac{r}{k} (\alpha_0 \alpha + \beta_0 \beta + \gamma_0 \gamma) \tag{10.10}$$

This is of the form

$$\mathbf{k} \cdot \mathbf{r} = f(\alpha, \beta) \tag{10.11}$$

In the vicinity of $\alpha = \alpha_0$, $\beta = \beta_0$ we can use a Taylor series expansion for $f(\alpha, \beta)$ as

$$f(\alpha, \beta) \cong f(\alpha_0, \beta_0) + (\alpha - \alpha_0) f_\alpha(\alpha_0, \beta_0)$$

$$+ (\beta - \beta_0) f_\beta(\alpha_0, \beta_0) + \frac{1}{2}(\alpha - \alpha_0)^2 f_{\alpha\alpha}(\alpha_0, \beta_0)$$

$$+ (\alpha - \alpha_0)(\beta - \beta_0) f_{\alpha\beta}(\alpha_0, \beta_0) + \frac{1}{2}(\beta - \beta_0)^2 f_{\beta\beta}(\alpha_0, \beta_0) \tag{10.12}$$

The first derivatives are zero at α_0, β_0. In addition,

$$f(\alpha_0, \beta_0) = kr \tag{10.13}$$

so the constant term in the Taylor series expansion is just the normal spherical wave phase. It can be easily shown that the second partial derivatives are given as

$$f_{\alpha\alpha}(\alpha_0, \beta_0) = -\frac{r}{k} \frac{k^2 - \beta_0^2}{k^2 - \alpha_0^2 - \beta_0^2} \tag{10.14}$$

$$f_{\alpha\beta}(\alpha_0, \beta_0) = -\frac{r}{k} \frac{\alpha_0\beta_0}{k^2 - \alpha_0^2 - \beta_0^2} \tag{10.15}$$

$$f_{\beta\beta}(\alpha_0, \beta_0) = -\frac{r}{k} \frac{k^2 - \alpha_0^2}{k^2 - \alpha_0^2 - \beta_0^2} \tag{10.16}$$

Thus, the exponential term in (10.7) becomes

$$e^{-j\mathbf{k}\cdot\mathbf{r}} = e^{-jf(\alpha,\beta)} = e^{-jf(\alpha_0,\beta_0)}$$

$$\cdot e^{-j\left[\frac{1}{2}(\alpha - \alpha_0)^2 f_{\alpha\alpha}(\alpha_0,\beta_0) + (\alpha - \alpha_0)(\beta - \beta_0)f_{\alpha\beta}(\alpha_0,\beta_0) + \frac{1}{2}(\beta - \beta_0)^2 f_{\beta\beta}(\alpha_0,\beta_0)\right]} \tag{10.17}$$

$$= e^{-jkr} e^{j[A(\alpha - \alpha_0)^2 + 2B(\alpha - \alpha_0)(\beta - \beta_0) + C(\beta - \beta_0)^2]}$$

where

$$A = \frac{r}{2k} \frac{k^2 - \beta_0^2}{k^2 - \alpha_0^2 - \beta_0^2} \tag{10.18}$$

$$B = \frac{r}{2k} \frac{\alpha_0\beta_0}{k^2 - \alpha_0^2 - \beta_0^2} \tag{10.19}$$

$$C = \frac{r}{2k} \frac{k^2 - \alpha_0^2}{k^2 - \alpha_0^2 - \beta_0^2} \tag{10.20}$$

So, (10.7) becomes

$$\mathbf{E}(\mathbf{r}) \cong e^{-jkr} \iint_{-\infty}^{\infty} \mathbf{E}(\bar{\alpha}, \bar{\beta}) \, e^{j[A\bar{\alpha}^2 + 2B\bar{\alpha}\bar{\beta} + C\bar{\beta}^2]} \, d\bar{\alpha} \, d\bar{\beta} \tag{10.21}$$

where

$$\bar{\alpha} = \alpha - \alpha_0 \tag{10.22}$$

$$\bar{\beta} = \beta - \beta_0 \tag{10.23}$$

The exponent is a quadratic form given by

$$Q(\bar{\alpha}, \bar{\beta}) = (\bar{\alpha}, \bar{\beta})[M]\begin{pmatrix}\bar{\alpha}\\\bar{\beta}\end{pmatrix} \tag{10.24}$$

where

$$[M] = \frac{r}{2k(k^2 - \alpha_0^2 - \beta_0^2)}\begin{bmatrix} k^2 - \beta_0^2 & \alpha_0\beta_0 \\ \alpha_0\beta_0 & k^2 - \alpha_0^2 \end{bmatrix} \qquad (10.25)$$

The matrix $[M]$ may be diagonalized by an appropriate rotation of coordinates in $\bar{\alpha}$, $\bar{\beta}$ space. This is equivalent to a transformation from an $\hat{x}$, $\hat{y}$-based coordinate system to a $\hat{\theta}$, $\hat{\phi}$-based (i.e., ray-based) coordinate system. We will denote

$$\mathbf{x} = \begin{pmatrix} \bar{\alpha} \\ \bar{\beta} \end{pmatrix} \text{ and } \mathbf{x}^T = (\bar{\alpha}, \bar{\beta}) \qquad (10.26)$$

The quadratic form can then be rewritten as

$$Q(\mathbf{x}) = \mathbf{x}^T[M]\mathbf{x} \qquad (10.27)$$

We now define a new vector, $\mathbf{u}$, such that

$$\mathbf{u} = \begin{pmatrix} u_1 \\ u_2 \end{pmatrix} \qquad (10.28)$$

and,

$$\mathbf{x} = [R]\mathbf{u}, \mathbf{x}^T = \mathbf{u}^T[R]^T \qquad (10.29)$$

Now the quadratic form becomes

$$Q(\mathbf{u}) = \mathbf{u}^T[R^TMR]\mathbf{u} \qquad (10.30)$$

If we select $[R]$ to be

$$[R] = \frac{1}{(\alpha_0^2 + \beta_0^2)^{1/2}}\begin{bmatrix} \alpha_0 & -\beta_0 \\ \beta_0 & \alpha_0 \end{bmatrix} \qquad (10.31)$$

then

$$[R^TMR] = \frac{r}{2k}\begin{bmatrix} \dfrac{k^2}{k^2 - \alpha_0^2 - \beta_0^2} & 0 \\ 0 & 1 \end{bmatrix} \qquad (10.32)$$

Since the transformation defined by $[R]$ is unitary (i.e., it merely represents a rotation of coordinate axes), we have

$$d\bar{\alpha}\ d\bar{\beta} = du_1\ du_2 \tag{10.33}$$

Hence, (10.21) becomes

$$\mathbf{E}(\mathbf{r}) \cong e^{-jkr} \int\!\!\int_{-\infty}^{\infty} \mathbf{E}(u_1, u_2)\ e^{jA'u_1^2}\ e^{jB'u_2^2}\ du_1\ du_2 \tag{10.34}$$

where

$$A' = \frac{kr}{2}\ (k^2 - \alpha_0^2 - \beta_0^2)^{-1} \tag{10.35}$$

$$B' = \frac{r}{2k} \tag{10.36}$$

The integral in (10.34) becomes

$$\mathbf{E}(\mathbf{r}) \cong \mathbf{E}(\alpha_0, \beta_0)\ e^{-jkr} \int_{-\infty}^{\infty} e^{jA'u_1^2}\ du_1 \int_{-\infty}^{\infty} e^{jB'u_2^2}\ du_2 \tag{10.37}$$

The integrals in (10.37) are easily evaluated using the result

$$\int_{-\infty}^{\infty} e^{jAx^2}\ dx = \sqrt{\frac{\pi}{A}}\ e^{j\pi/4} \tag{10.38}$$

So, (10.37) becomes

$$\mathbf{E}(\mathbf{r}) \cong j2\pi\ \frac{e^{-jkr}}{r}\ \mathbf{E}(\alpha_0, \beta_0)\ \sqrt{k^2 - \alpha_0^2 - \beta_0^2} \tag{10.39}$$

Equation (10.39) expresses the tangential component of the far-zone electric field in terms of the tangential components of the field at $z = 0$. Gaussian law, which states that

$$\nabla \cdot \mathbf{E} = 0 \tag{10.40}$$

can then be used to calculate the z-component of the fields.

The real beauty of calculating the radiated electric field via (10.39), rather than integrating the currents on the reflector, is that we may now evaluate the quantity $\mathbf{E}\,(\alpha,\ \beta)$ using any of the techniques in Chapters 4–7, specialized to the case of a planar integration surface. This is where the real speed and efficiency of the aperture integration method comes from.

10.2 CALCULATION OF APERTURE ELECTRIC FIELDS VIA GEOMETRICAL OPTICS

Now that we know how to calculate radiated far fields in terms of the aperture electric field, we now turn to the problem of how to calculate the aperture electric field of a reflector antenna system. Consider Figures 10.2 and 10.3 which illustrate symmetrical and offset reflectors, along with their associated integration aperture planes. The fields over the portion of the aperture plane eclipsed by the reflector is assumed equal to the geometrical optics field. The field elsewhere is assumed zero.

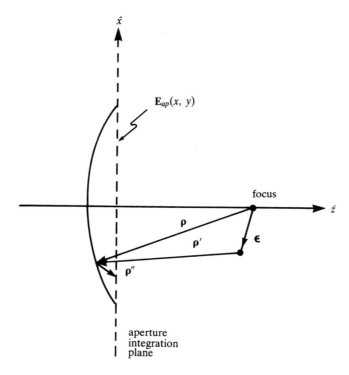

Figure 10.2 Symmetrical reflector antenna, with integration aperture.

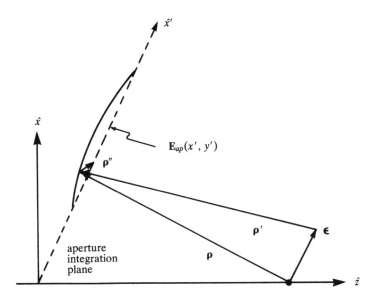

Figure 10.3 Offset reflector antenna, with integration aperture.

Since the projected aperture of both reflectors onto the x-y plane is circular, it is perhaps best to integrate the aperture fields with respect to a polar coordinate system in the x-y plane. We seek to calculate the magnitude, phase, and polarization of the geometrical optics electric field over the integration aperture of each reflector antenna type.

Using the following notation,

ρ' = vector from feed to reflector surface
ρ'' = vector from reflector surface to integration aperture

we would first select a point on the aperture integration plane where we wish to calculate the GO electric field. We then calculate the reflection point on the reflector surface as that point which minimizes the quantity

$$|\rho'| + |\rho''|$$

Once the reflection point is known, the aperture field phase is given as

$$\phi = e^{-jk\{|\rho'| + |\rho''|\}} \tag{10.41}$$

The surface normal at the reflection point now can be calculated. This allows the reflected field polarization to be written as

$$\mathbf{E}^{\text{refl}} = -\mathbf{E}^{\text{inc}} + 2[\hat{n} \cdot \mathbf{E}^{\text{inc}}] \, \mathbf{E}^{\text{inc}} \tag{10.42}$$

The only thing left now is to calculate the reflected field magnitude.

Calculating the magnitude of the reflected field is an interesting exercise involving the geometrical properties of the incident and reflected wavefronts as well as those of the curved reflecting surface. The local curvature of the reflecting surface causes the local curvature of the incident electric field to be transformed upon reflection. This of course is a well known phenomenon which forms the basis of Cassegrain antenna design. In this case however we study this phenomenon as it pertains to localized quadratic patches on smooth 2 dimensional surfaces. Our approach will follow that of [1].

To begin, consider Figure 10.4 which illustrates a local patch of a curved phasefront. Assume that the unit vectors $\hat{x}_1$, $\hat{x}_2$ lie in the plane tangent to the phasefront at 0 and that they lie in the principal planes of curvature (the subject of curvature will be discussed at length shortly). So, if the phasefront is normal to $\hat{z}$ at 0, the phase over the $z = 0$ plane is given by the Taylor series

$$\phi\,(y_1, y_2) = \phi\,(0, 0) + \frac{1}{2}y_1^2\,\phi_{y_1 y_1}\,(0, 0)$$

$$+ \ y_1 y_2 \,\phi_{y_1 y_2}\,(0, 0) + \frac{1}{2}\,y_2^2 \phi_{y_2 y_2}\,(0, 0) \tag{10.43}$$

where y_1, y_2 are variables in the direction of unit vectors $\hat{y}_1$, $\hat{y}_2$ and $\hat{y}_1$, $\hat{y}_2$ are any set of orthogonal unit vectors in the $\hat{x}_1$, $\hat{x}_2$ plane.

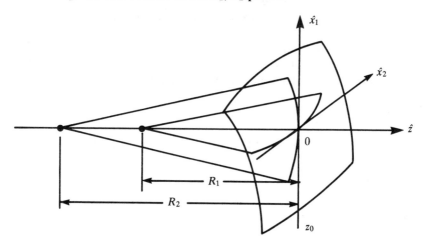

Figure 10.4 Local patch of wavefront surface.

The two basis sets are related by the rotation operator:

$$\hat{y}_1 = \cos \alpha \, \hat{x}_1 + \sin \alpha \, \hat{x}_2 \tag{10.44}$$

$$\hat{y}_2 = -\sin \alpha \, \hat{x}_1 + \cos \alpha \, \hat{x}_2 \tag{10.45}$$

The variables are related in a similar way

$$\begin{bmatrix} y_1 \\ y_1 \end{bmatrix} = \begin{bmatrix} \cos \alpha & \sin \alpha \\ -\sin \alpha & \cos \alpha \end{bmatrix} \begin{bmatrix} x_1 \\ x_2 \end{bmatrix} \tag{10.46}$$

The angle, α, is determined from (10.43) as follows. Rewrite (10.43) as

$$\phi(y_1, y_2) - \phi(0, 0) = Q(y_1, y_2)$$

$$= \frac{1}{2} (y_1, y_2) \begin{bmatrix} \phi_{y_1 y_1} & \phi_{y_1 y_2} \\ \phi_{y_1 y_2} & \phi_{y_2 y_2} \end{bmatrix} \begin{bmatrix} y_1 \\ y_2 \end{bmatrix}$$

$$= \frac{1}{2} \mathbf{y}^T [M] \mathbf{y} \tag{10.47}$$

which is merely the same type of quadratic form encountered earlier in this chapter. This quadratic form can be diagonalized using a simple rotation operator of the type in (10.46). So, we write

$$\mathbf{y} = [R]\mathbf{x}, \qquad \mathbf{y}^T = \mathbf{x}^T [R]^T \tag{10.48}$$

Substituting (10.48) into (10.47) gives

$$Q(x_1, x_2) = \frac{1}{2} \mathbf{x}^T [R^T M R] \, \mathbf{x} \tag{10.49}$$

Performing the matrix operations in (10.49) shows that the diagonal elements of $R^T M R$ are zero when

$$\tan 2\alpha = \frac{2\phi_{y_1 y_2}}{\phi_{y_2 y_2} - \phi_{y_1 y_1}} \tag{10.50}$$

or

$$\alpha = \frac{1}{2} \tan^{-1} \frac{2\phi_{y_1 y_2}}{\phi_{y_2 y_2} - \phi_{y_1 y_1}} \tag{10.51}$$

So, knowing the equation of the phase front, it's easy to calculate the principal planes of the phase front. The diagonal terms of $R^T M R$ are

$$\phi_{x_1 x_1} = \phi_{y_1 y_1} \cos^2 \alpha + \phi_{y_2 y_2} \sin^2 \alpha - 2\phi_{y_1 y_2} \sin \alpha \cos \alpha \tag{10.52}$$

and

$$\phi_{x_2 x_2} = \phi_{y_1 y_1} \sin^2 \alpha + \phi_{y_2 y_2} \cos^2 \alpha + 2\phi_{y_1 y_2} \sin \alpha \cos \alpha \tag{10.53}$$

Also

$$\phi_{x_1 x_1} = \frac{1}{R_1} \text{ and } \phi_{x_2 x_2} = \frac{1}{R_2} \tag{10.54}$$

where R_1, R_2 are the principal radii of curvature of the phasefront. The quantity R_1, R_2 is often referred to as *Gaussian curvature* and is equal to the inverse of the determinant of M or $R^T M R$.

The matrix $M' = R^T M R$ is a function of z. Thus,

$$M'(z) = \begin{bmatrix} \dfrac{1}{R_1 + (z - z_0)} & 0 \\[2ex] 0 & \dfrac{1}{R_2 + (z - z_0)} \end{bmatrix} \tag{10.55}$$

The scalar amplitude and phase of the wave having the phase front shown in Figure 10.4 can be written as

$$u(y_1, y_2, z) = u(0, 0, z_0) \sqrt{\frac{\det M'(z)}{\det M'(z_0)}} \, e^{-jkz} \, e^{-j\frac{k}{2} y^T M(z) y} \tag{10.56}$$

Note: when $M'(z) = M'(z_0) = 0$ (perfectly collimated plane-wave phase front), merely set the quotient in (10.56) equal to 1.

We now consider what happens when a curved wavefront impinges on a smooth conducting curved surface. Consider Figure 10.5 which illustrates the sit-

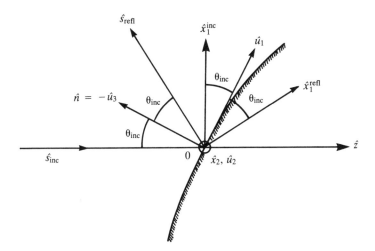

Figure 10.5 Scattering of a curved phasefront by a curved conducting surface.

uation schematically. Since the surface is assumed smooth, we can write the equation for the surface in terms of a Taylor series about 0:

$$u_3(u_1, u_2) = \frac{1}{2} (u_1, u_2) \begin{bmatrix} \partial^2 u_3/\partial u_1^2 & \partial^2 u_3/\partial u_1 \partial u_2 \\ \partial^2 u_3/\partial u_1 \partial u_2 & \partial^2 u_3/\partial u_2^2 \end{bmatrix} \begin{bmatrix} u_1 \\ u_2 \end{bmatrix} \tag{10.57}$$

By 10.57, a vector to the surface is

$$\mathbf{r}(u_1, u_2) = u_1 \hat{u}_1 + u_2 \hat{u}_2 + \left[\frac{1}{2} \mathbf{u}^T S \mathbf{u} \right] \hat{u}_3$$

$$= \mathbf{u} + \left[\frac{1}{2} \mathbf{u}^T S \mathbf{u} \right] \hat{u}_3 \tag{10.58}$$

Likewise, the incident field phase is

$$\phi(x_1, x_2, z) = kz + k \left(\frac{1}{2} \mathbf{x}^T M'(z) \mathbf{x} \right) \tag{10.59}$$

by (10.56).

Dotting (10.58) with $\hat{z}$ gives

$$z_{\text{surf}} = \mathbf{u} \cdot \hat{z} + \frac{1}{2} (\mathbf{u}^T S \mathbf{u}) (\hat{u}_3 \cdot \hat{z}) \tag{10.60}$$

Substituting (10.60) into (10.59) gives

$$\phi_{\text{inc}} = k\{\mathbf{u} \cdot \hat{z} + \frac{1}{2}(\mathbf{u}^T S \mathbf{u})(\hat{u}_3 \cdot \hat{z})$$

$$+ \frac{1}{2}\mathbf{x}^T M'(z_{\text{surf}})\,\mathbf{x}\} \tag{10.61}$$

for the incident field phase (neglecting the factor of $-j$) at x_1, x_2, z_{surf}.

Next we express all quantities in (10.61) in terms of the tangent plane co-ordinates u_1, u_2 rather than the incident field (ray-based) coordinates x_1, x_2. So, since the projections of $\hat{x}_1$, $\hat{x}_2$ onto the $\mathbf{u}$ plane are

$$\hat{x}_1 = (\hat{x}_1 \cdot \hat{u}_1)\,\hat{u}_1 + (\hat{x}_1 \cdot \hat{u}_2)\,\hat{u}_2 \tag{10.62}$$

$$\hat{x}_2 = (\hat{x}_2 \cdot \hat{u}_1)\hat{u}_1 + (\hat{x}_2 \cdot \hat{u}_2)\,\hat{u}_2 \tag{10.63}$$

we have

$$\begin{pmatrix} x_1 \\ x_2 \end{pmatrix} = \begin{pmatrix} \hat{x}_1 \cdot \hat{u}_1 & \hat{x}_1 \cdot \hat{u}_2 \\ \hat{x}_2 \cdot \hat{u}_1 & \hat{x}_2 \cdot \hat{u}_2 \end{pmatrix} \begin{pmatrix} u_1 \\ u_2 \end{pmatrix} \tag{10.64}$$

or

$$\mathbf{x} = R\,\mathbf{u}\,;\,\mathbf{x}^T = \mathbf{u}^T R^t \tag{10.65}$$

Substituting (10.65) into (10.61) gives

$$\phi_{\text{inc}} = k\{\mathbf{u} \cdot \hat{z} + \frac{1}{2}(\mathbf{u}^T S\mathbf{u})(\hat{u}_3 \cdot \hat{z})$$

$$+ \frac{1}{2}\mathbf{u}^T(R^T M'R)\mathbf{u}\} \tag{10.66}$$

for the incident field phase at some point on the surface defined by u_1, u_2. Since $\hat{z} = \hat{s}_{\text{inc}}$, we can rewrite (10.66) as

$$\phi_{\text{inc}} = k\{\mathbf{u} \cdot \hat{s}_{\text{inc}} + \frac{1}{2}\mathbf{u}^T[(\hat{u}_3 \cdot \hat{s}_{\text{inc}})\,S + R^T M'_{\text{inc}}R]\,\mathbf{u}\} \tag{10.67}$$

Similarly, we have for the reflected field

$$\phi_{\text{refl}} = k\{\mathbf{u} \cdot \hat{s}_{\text{refl}} + \frac{1}{2}\mathbf{u}^T[(\hat{u}_3 \cdot \hat{s}_{\text{refl}}) \, S + R^T M'_{\text{refl}} R] \, \mathbf{u}\} \tag{10.68}$$

where (10.68) has been written under the assumption that vectors $\hat{x}_1$, $\hat{x}_2$ for the reflected field have been chosen such that (10.64) holds for reflected as well as incident fields (i.e., we already assumed Snell's law).

We now equate (10.67) and (10.68) to get

$$\mathbf{u} \cdot \hat{s}_{\text{inc}} = \mathbf{u} \cdot \hat{s}_{\text{refl}} \quad \text{(Snell's law)} \tag{10.69}$$

and

$$(\hat{u}_3 \cdot \hat{s}_{\text{inc}}) \, S + R^T M'_{\text{inc}} R$$

$$= (\hat{u}_3 \cdot \hat{s}_{\text{refl}}) \, S + R^T M'_{\text{refl}} R \tag{10.70}$$

or, since $\hat{u}_3 \cdot \hat{s}_{\text{inc}} = -\hat{u}_3 \cdot \hat{s}_{\text{refl}}$,

$$R^T M'_{\text{refl}} \, R = 2(\hat{u}_3 \cdot \hat{s}_{\text{inc}}) S + R^T M'_{\text{inc}} R \tag{10.71}$$

$$M'_{\text{refl}} = 2(\hat{u}_3 \cdot \hat{s}_{\text{inc}}) \, R^{T^{-1}} S R^{-1} + M'_{\text{inc}} \tag{10.72}$$

So, (10.72) gives the curvature matrix for the reflected wavefront (at 0) in terms of the curvature matrix, S, for the surface (at 0) and the curvature matrix, M'_{inc}, for the incident field (at 0).

The curvature matrix of the reflected wavefront, M'refl, can now be diagonalized to obtain the two principal radii of curvature R_1, R_2 of the reflected wavefront at 0. The reflected field can then be calculated as

$$\mathbf{E}^{\text{refl}}(s_r) = \mathbf{E}^{\text{refl}}(s_r = 0) \, e^{-j\phi_{\text{inc}}} e^{-jks_r}$$

$$\cdot \sqrt{\frac{R_1 R_2}{(R_1 + s_r)(R_2 + s_r)}} \tag{10.73}$$

where

$$s_r = \text{variable in the direction of } \hat{s}_{\text{refl}} \, (s_r = 0 \text{ at surface})$$
$$-\phi_{\text{inc}} = \text{incident field phase at 0}$$
$$R_1, R_2 = \text{principal radii of curvature of the reflected wavefront at 0}$$

Note: if $R_1, R_2 = \infty$ (plane wave), the quotient in (10.73) equals unity. The M, S matrices are all positive definite.

REFERENCES

1. Deschamps, G.A., "Ray Techniques in Electromagnetics," *Proc. IEEE,* Vol. 60, pp. 1022–1035, September 1972.

Suggestions for Additional Reading

1. Yaghjian, A.D., "Equivalence of Surface Current and Aperture Field Integrations for Reflector Antennas," *IEEE Trans. Antennas Propagat.,* Vol. AP-32, pp. 1355–1358, December 1984.
2. Rahmat-Samii, Y., "A Comparison Between GO/Aperture Field and Physical Optics Methods for Offset Reflectors," *IEEE Trans. Antennas Propagat.,* Vol. AP-32, pp. 301–306, March, 1984.
3. Matsuura, H. and K. Hongo, "Comparison of Induced Current and Aperture Field Integrations for an Offset Parabolic Reflector," *IEEE Trans. Antennas Propagat.,* Vol. AP-35, pp. 101–105, January 1987.
4. Hongo K. and H. Matsuura, "Simplified Technique for Evaluating the Radiation Integrals," *IEEE Trans. Antennas Propagat.,* Vol. AP-34, pp. 732–737, May 1986.

Appendix A
Surface Normals and Area Elements

A little differential geometry is necessary in reflector antenna analysis. Two quantities that repeatedly arise in this connection are the vector normal to the reflector, $\hat{n}$, and the elemental surface area, ds'. This Appendix shows how these quantities are calculated from the equation of the reflector surface.

Consider Figure A.1 which schematically shows an offset reflector. Reflector surfaces are generally described in one of two ways: parametrically or implicitly. In the parametric description, the 3-dimensional position vector, $\boldsymbol{\rho}$, describing the surface is expressed as a function of two scalar variables. That is, $\boldsymbol{\rho} = \boldsymbol{\rho}(u, v)$. In the implicit form, the surface is described in terms of a single scalar function of the three coordinate variables. In rectangular coordinates, this would take the form $f(x, y, z) = 0$.

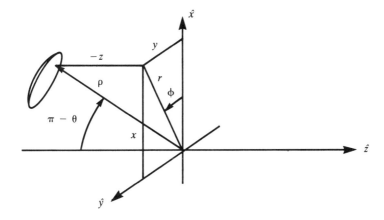

Figure A.1 Schematic diagram of an offset reflector antenna showing rectangular, cylindrical, and spherical coordinates.

In the case of a paraboloid of revolution, a parametric form in terms of spherical coordinates would be

$$\rho\,(\theta,\,\phi) = \left[\frac{2F}{1 - \cos\theta}\right](\hat{x}\,\sin\theta\,\cos\phi + \hat{y}\,\sin\theta\,\sin\phi + \hat{z}\,\cos\theta) \qquad \text{(A.1)}$$

where the origin is assumed located at the focal point of the paraboloid. In rectangular coordinates we would have

$$\rho\,(x,\,y) = x\,\hat{x} + y\,\hat{y} + \left[\frac{x^2 + y^2}{4F} - F\right]\hat{z} \qquad \text{(A.2)}$$

where, again, we assume the origin is located at the focal point.

An implicit form in terms of spherical coordinates would be

$$\rho - \frac{2F}{1 - \cos\theta} = 0 \qquad \text{(A.3)}$$

which is of the form $f(\rho,\,\theta,\,\phi) = 0$. An implicit form in terms of cylindrical coordinates would be

$$z - \frac{r^2}{4F} + F = 0 \qquad \text{(A.4)}$$

In the parametric description, the normal vector is given as

$$\hat{n} = \frac{\rho_u \times \rho_v}{|\rho_u \times \rho_v|} \qquad \text{(A.5)}$$

where the the subscripts u, v denote partial differentiation with respect to the two parameters. The two vectors ρ_u, ρ_v lie in the tangent plane to the surface, hence their cross product is normal to the tangent plane.

In implicit form,

$$\hat{n} = \frac{\nabla f}{|\nabla f|} \qquad \text{(A.6)}$$

where $f(u,\,v,\,w) = 0$ defines the surface. The element of surface area can be found from the parametric form as

$$ds' = |\rho_u \times \rho_v|\,du\,dv \qquad \text{(A.7)}$$

An easy alternative method also exists for calculating differential surface area. Consider Figure A.2, which illustrates a paraboloid of revolution. The element of surface area on the reflector surface can be calculated by merely taking the element of surface area of the (θ, ϕ) coordinate patch and then dividing by the cosine of the angle between the reflector normal and the normal to the (θ, ϕ) coordinate patch. For example, the area element in (θ, ϕ) coordinates is

$$ds'_{\theta,\phi} = \rho^2 \sin \theta \, d\theta \, d\phi \tag{A.8}$$

Now, since

$$\hat{n}_p \cdot \hat{n}_s = \cos \theta/2 \tag{A.9}$$

we have

$$ds' = \rho^2 \sec \frac{\theta}{2} \sin \theta \, d\theta \, d\phi \tag{A.10}$$

as the elemental area on the paraboloidal surface.

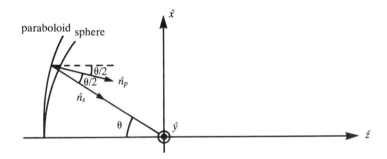

Figure A.2 On the computation of ds' in spherical coordinates.

In cylindrical coordinates, the differential area element in the x, y plane is given as

$$ds'_{r,\phi} = r \, dr \, d\phi \tag{A.11}$$

In this case, the normal to the r, ϕ coordinate patch is the z axis. So,

$$\hat{n}_p \cdot \hat{z} = \cos \frac{\theta}{2} = \cos \alpha \tag{A.12}$$

where

$$\alpha = \tan^{-1} \frac{dz}{dr} \tag{A.13}$$

Since

$$z = \frac{r^2}{4F} - F \tag{A.14}$$

we have

$$\tan \alpha = \frac{r}{2F} \tag{A.15}$$

and

$$\cos \alpha = \frac{1}{\sqrt{1 + (r/2F)^2}} \tag{A.16}$$

Thus,

$$ds' = \sqrt{1 + (r/2F)^2} \, r \, dr \, d\phi \tag{A.17}$$

Appendix B
Subreflector Analysis Techniques

The physical optics reflector analysis techniques described in the text apply primarily to large, focused (or nearly focused) reflectors. In the dual-reflector configuration, however, a small unfocused reflector is also present and must be analyzed. This subreflector—in contrast to the main reflector, which may produce a narrow pencil beam only fractions of a degree in extent—will produce a broad sector beam occupying most of the forward hemisphere.

The pattern differences of the two reflectors stem directly from the scattering mechanisms of each. The main reflector is a "diffraction limited" device, in that diffraction is the phenomenon which governs the far-field pattern characteristics. Geometrical optics does not account for any of the main reflector far-field characteristics (even though it may quite accurately predict the aperture fields). In the case of the subreflector, however, diffraction is a relatively minor phenomenon compared to the geometrical-optics ray contribution (see Figures B.1 and B.2).

The scatter pattern of the subreflector may be calculated in a number of ways. One way is to simply integrate the physical optics currents on the reflector [1–3]. This procedure is quite time consuming, especially since a full contour pattern over a very broad angular sector must be calculated in order to get the main reflector illumination function. So what we need is a very quick way of calculating the subreflector radiation pattern. Fortunately, extreme accuracy is not a predominant concern for this calculation. The fields incident on the main reflector may be in error by a surprising degree without seriously affecting the accuracy of the secondary pencil-beam pattern (it is a general property of electromagnetic fields that they are stationary with respect to their source distributions).

One approach to calculating the subreflector pattern takes advantage of the fact that the geometrical optics contribution dominates and the diffracted field contribution represents a second-order "pattern correction" type of effect. This is the GO/GTD (geometrical theory of diffraction) approach [4–5]. In this case, the diffracted field contribution is approximated by a ray-optics field (analogous to the geometrical optics reflected field) which is regarded as originating at a point (or

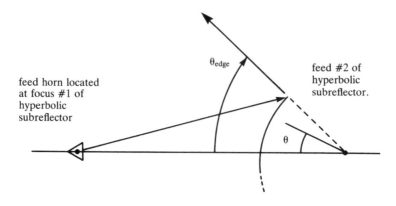

Figure B.1 Schematic diagram of the feed horn and subreflector geometry.

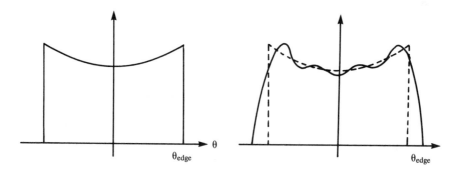

a. geometrical optics contribution   b. geometrical optics contribution plus diffraction effects

Figure B.2 Typical subreflector scatter patterns (for the geometry shown in Figure B.1).

points) located on the rim of the subreflector. The point(s) on the rim where this edge-diffracted ray originates will be determined by logic similar to that used in locating the specular reflection point on a surface.

The geometrical optics reflected field may be calculated using the methods of Chapter 10. The geometrical optics specular reflection point will be calculated slightly differently in this case, however. With reference to Figure B.3, we may require that $\hat{s}_{\text{refl}}$, $\hat{R}$ be colinear:

$$|\hat{s}_{\text{refl}} \times \hat{R}| = 0 \tag{B.1}$$

Now, since

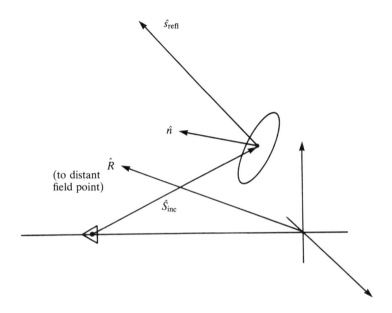

Figure B.3 On the calculation of the specular reflection point on the surface of the subreflector.

$$\hat{s}_{\text{refl}} = \hat{s}_{\text{inc}} - 2(\hat{n} \cdot \hat{s}_{\text{inc}})\hat{n} \tag{B.2}$$

(B.1) becomes

$$\left|[\hat{s}_{\text{inc}} - 2(\hat{n} \cdot \hat{s}_{\text{inc}})\hat{n}] \times \hat{R}\right| = 0 \tag{B.3}$$

A search routine can be used to find the point on the subreflector surface where the magnitude of the cross product is zero.

The geometrical optics solution alone is often accurate enough for calculating far-zone fields scattered by the main reflector (in fact, the geometrical optics contribution was all that was used in the dual-reflector design process from Chapter 9). However, in many cases it is desirable to calculate the subreflector pattern to greater accuracy. To do this, we add in the diffracted field contribution. This is done using a ray-based technique known as the *geometrical theory of diffraction* (GTD) [6–7].

A detailed derivation of the GTD concept as an asymptotic solution to the problem of electromagnetic scattering by a conducting half-plane would be beyond the scope of this book. The discussion herein will be confined primarily to explaining how GTD works and how it can be applied to the problem of calculating fields diffracted by a subreflector.

To this end, consider Figure B.4 which shows a ray incident on the edge of some surface. By [7], the edge diffracted field is given by

$$\begin{bmatrix} E_\beta^d \\ E_\phi^d \end{bmatrix} = - \begin{bmatrix} E_\beta^i \cdot D_s \\ E_\phi^i \cdot D_h \end{bmatrix} \sqrt{\frac{R_c}{s_d(R_c + s_d)}} \, e^{-jks_d} \tag{B.4}$$

where

E_β^d, E_ϕ^d represent the edge-diffracted fields in the $\hat{\beta}_d$, $\hat{\phi}_d$ directions, respectively

E_β^i, E_ϕ^i represent the incident fields in the $\hat{\beta}_i$, $\hat{\phi}_i$ directions, respectively

D_s, D_h represent the ("soft" and "hard") diffraction coefficients which apply to the β, ϕ components of electric field, respectively

R_c represents the caustic distance (equal to infinity in the case of a straight edge). This will be discussed in greater detail below.

s_d represents the distance along the ray $\hat{s}_d$.

Note that the angle, ϕ_d, of the diffracted ray is arbitrary. That is, the diffracted rays radiate in a cone about the vector, $\hat{t}$, tangent to the edge. In this sense, the

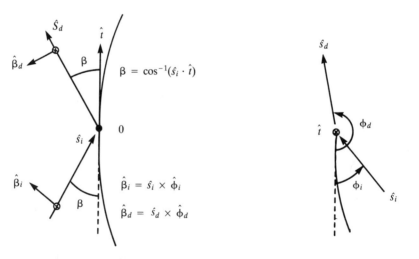

a. viewing in the plane defined by $\hat{s}_i$, $\hat{\beta}_i$.   b. viewing along the tangent vector, $\hat{t}$.

Figure B.4 Geometry of edge diffraction problem (*note:* all vectors are projected onto their various viewing planes).

edge radiates the diffracted field in much the same way that an infinite linear current filament would radiate energy. The primary difference is that the edge radiates both polarizations and that the edge does not radiate uniformly in all directions, as a current filament would.

The diffraction coefficients are obtained by asymptotically evaluating the solution to the half-plane problem. This is a very involved procedure, which results in the following expressions for the soft and hard diffraction coefficients [7]:

$$
D_{\substack{s\\h}}(\phi_d, \phi_i; \beta) = - \frac{e^{-j\frac{\pi}{4}}}{2\sqrt{2\pi k}\,\sin\beta}
$$

$$
\cdot \left\{ \frac{F[kL^i\,a(\phi_d - \phi_i)]}{\cos[(\phi_d - \phi_i)/2]} \; \substack{-\\+} \; \frac{F[kL^d\,a(\phi_d + \phi_i)]}{\cos[(\phi_d + \phi_i)/2]} \right\} \tag{B.5}
$$

where

$$
a(x) = 2\cos^2(x/2) \tag{B.6}
$$

$$
L^i = \frac{s_d\,R_{\text{inc}}\,\sin^2\beta}{(R_{\text{inc}} + s_d)} \tag{B.7}
$$

$$
L^d = \frac{s_d\,(R_c + s_d)\,R_1^d\,R_2^d\,\sin^2\beta}{R_c\,(R_1^d + s_d)\,(R_2^d + s_d)} \tag{B.8}
$$

and

$$
F(x) = 2j\,\sqrt{x}\,e^{jx}\int_{\sqrt{x}}^{\infty} e^{-j\tau^2}\,d\tau \tag{B.9}
$$

In the equations above,

R_{inc} = radius of curvature of the incident wave (assumed to be the same in both principal planes)

R_1^d, R_2^d = principal radii of curvature of the diffracted wavefront

R_c = caustic distance

If the caustic distance R_c is positive, there is no caustic along the diffracted ray path (i.e., the diffracted wave is locally divergent). If R_c is negative, the caustic lies between the edge-diffraction point, O, and the far-field observation point (i.e., the diffracted wave is locally convergent near O). In this latter case, a phase shift of $+\pi/2$ is naturally introduced in the radical term in (B.4) as s_d passes the point, $s_d = |R_c|$. As $s_d \to \infty$,

$$L^i \to R_{\text{inc}} \sin^2 \beta \tag{B.10}$$

$$L^d \to \frac{R_1^d R_2^d}{R_c} \sin^2 \beta \tag{B.11}$$

It still remains to determine the location of the edge diffraction points as well as the caustic distance, R_c. With reference to Figure B.4, the edge diffraction points are located by the equation

$$\hat{t} \cdot \hat{s}_i = \hat{t} \cdot \hat{s}_d \tag{B.12}$$

It is a relatively easy matter to search around the rim of the subreflector, locating any points that satisfy this criterion.

For certain observation angles, there may be points on the reflector rim which satisfy (B.12) which do not contribute to the diffracted field. These are points which are shadowed by the reflector surface, as indicated in Figure B.5. In this case, the observation point is in the $\phi = 0$ plane and the edge diffraction point at $\phi = 180°$ on the subreflector is shadowed as shown.

The caustic distance for the diffracted ray is calculated in much the same way that the caustic distances for reflection were calculated in Chapter 10. To begin, consider Figure B.6 which illustrates the problem of ray diffraction by a curved edge in a surface. In the figure,

$\hat{t}$ = unit vector tangent to edge curve

$\hat{n}$ = unit curvature vector for the edge curve (which may or may not lie in the surface tangent plane)

The equation for the edge curve near O is

$$\mathbf{r}(t) - \mathbf{r}_0 = t\,\hat{t} + \frac{1}{2}\frac{t^2}{\rho}\,\hat{n} \tag{B.13}$$

where ρ is the radius of curvature of the edge curve at 0. The phase of the incident (spherical) wave is

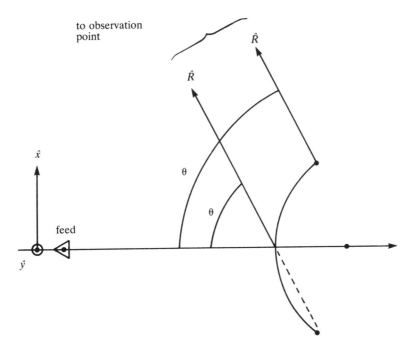

Figure B.5 On the shadowing effect for a symmetric hyperbolic subreflector.

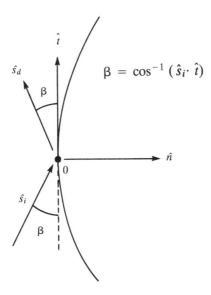

Figure B.6 On the calculation of the caustic distance for diffraction.

$$\psi_i(s_i, \mathbf{x}_i) = k \left(s_i + \frac{1}{2} \mathbf{x}_i^T Q_i \mathbf{x}_i \right) \tag{B.14}$$

where

$$Q_i = \frac{1}{R_i} \begin{bmatrix} 1 & 0 \\ 0 & 1 \end{bmatrix} \tag{B.15}$$

and $\mathbf{x}_i$ is some vector perpendicular to $\hat{s}_i$. Since

$$s_i = (\mathbf{r} - \mathbf{r}_0) \cdot \hat{s}_i \tag{B.16}$$

$$s_i(t) = t(\hat{t} \cdot \hat{s}_i) + \frac{1}{2} \frac{t^2}{\rho} (\hat{n} \cdot \hat{s}_i) \tag{B.17}$$

Hence,

$$\psi_i = k \left[t(\hat{t} \cdot \hat{s}_i) + \frac{1}{2} \frac{t^2}{\rho} (\hat{n} \cdot \hat{s}_i) + \frac{1}{2} \mathbf{x}_i^T Q_i \mathbf{x}_i \right] \tag{B.18}$$

$$\psi_d = k \left[t(\hat{t} \cdot \hat{s}_d) + \frac{1}{2} \frac{t^2}{\rho} (\hat{n} \cdot \hat{s}_d) + \frac{1}{2} \mathbf{x}_d^T Q_d \mathbf{x}_d \right] \tag{B.19}$$

Now let

$$\mathbf{x}_i = (\mathbf{x}_i \cdot \hat{t}) \hat{t} + (\mathbf{x}_i \cdot \hat{n}) \hat{n} = \mathbf{w} + (\mathbf{x}_i \cdot \hat{n}) \hat{n} \tag{B.20}$$

$$\mathbf{x}_d = (\mathbf{x}_d \cdot \hat{t}) \hat{t} + (\mathbf{x}_d \cdot \hat{n}) \hat{n} = \mathbf{w} + (\mathbf{x}_d \cdot \hat{n}) \hat{n} \tag{B.21}$$

By (B.13), if $\mathbf{r}' = \mathbf{r} - \mathbf{r}_0$ is some vector ranging from the point 0 to a point on the edge curve near 0, then the component of $\mathbf{r}'$ which is parallel to $\hat{n}$ is on the order of t^2, whereas the component parallel to $\hat{t}$ is on the order of t. This means that

$$\mathbf{x}_i^T Q_i \mathbf{x} = \mathbf{w}_i^T Q_i \mathbf{w}_i + \text{terms on the order of } t^3 \text{ and greater.} \tag{B.22}$$

Hence, to the second order, (B.18) and (B.19) become

$$\psi_i = k\left[t(\hat{t} \cdot \hat{s}_i) + \frac{1}{2}\frac{t^2}{\rho}(\hat{n} \cdot \hat{s}_i) + \frac{1}{2}\mathbf{w}_i^T Q_i \mathbf{w}_i \right] \qquad (B.23)$$

$$\psi_d = k\left[t(\hat{t} \cdot \hat{s}_d) + \frac{1}{2}\frac{t^2}{\rho}(\hat{n} \cdot \hat{s}_d) + \frac{1}{2}\mathbf{w}_d^T Q_d \mathbf{w}_d \right] \qquad (B.24)$$

Now, the vectors **w** may lie anywhere in the plane defined by the vector $\hat{t}$ and the vector $\hat{t} \times \hat{n}$. On the other hand, the continuity of phase boundary condition can only be enforced in the direction of $\hat{t}$ (i.e., along the edge curve). So, setting $\mathbf{w}_i = \mathbf{w}_d = t\hat{t}$ in (B.23) and (B.24) we get

$$\psi_i = k\left[t(\hat{t} \cdot \hat{s}_i) + \frac{1}{2}\frac{t^2}{\rho}(\hat{n} \cdot \hat{s}_i) + \frac{1}{2}t^2/R_i \right] \qquad (B.25)$$

$$\psi_d = k\left[t(\hat{t} \cdot \hat{s}_d) + \frac{1}{2}\frac{t^2}{\rho}(\hat{n} \cdot \hat{s}_d) + \frac{1}{2}t^2/R_c \right] \qquad (B.26)$$

to the order of t^3, where R_c is the caustic distance.

Note: in deriving (B.25) and (B.26), the vectors **w** were taken with respect to the basis set defined by $\hat{t}$ and $\hat{t} \times \hat{n}$, so that only the first component of these vectors is non-zero.

So, equating ψ_i, ψ_d in (B.25) and (B.26) gives

$$1. \ \hat{t} \cdot \hat{s}_i = \hat{t} \cdot \hat{s}_d \qquad (B.27)$$

$$2. \ \frac{1}{R_c} = \frac{1}{R_i} + \frac{1}{\rho \sin^2 \beta}\hat{n} \cdot (\hat{s}_i - \hat{s}_d) \qquad (B.28)$$

The last equation is our desired equation for the caustic distance.

Spherical Wave Theory

An alternate approach to the calculation of subreflector scatter patterns involves the use of spherical wave harmonics [8–12]. A brief outline of this approach is given below. First, consider Figure B.7. Spherical wave theory can be employed in two different ways in the calculation of the subreflector radiation pattern. These two situations are detailed below.

If it happens that the subreflector is located in the near field of the feed, then clearly the far-field radiation pattern of the feed will not adequately describe the

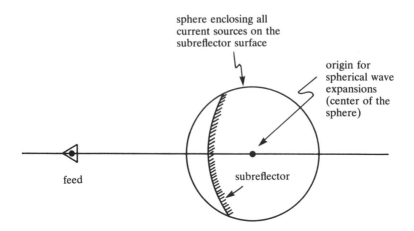

Figure B.7 On the spherical wave technique.

fields incident on the subreflector. In this case, the feed radiation pattern can be expanded in a series of spherical wave harmonics. This series will be valid every-where outside the smallest sphere (concentric to the sphere over which the feed pattern is matched to the spherical wave series) enclosing the feed. The coefficients of the various spherical wave harmonics are obtained using the orthogonality properties of the spherical mode functions (w.r.t. θ, φ). Once the coefficients of the spherical wave series are known, they can be used to calculate the feed pattern at the subreflector surface (assumed in the near-field of the feed). *Note:* to obtain the spherical wave expansion of the feed radiation pattern, the feed pattern is equated to the spherical wave series over a sphere of infinite radius, so that the radial functions can be replaced by their asymptotic forms.

Spherical wave techniques can also be used when the main reflector lies in the near-field of the subreflector, or whenever the GO/GTD method does not yield accurate enough values for the fields radiated by the subreflector. In this case, we would select some small sphere which just encloses the current sources on the subreflector (as shown in Figure B.7). The field radiated by the subreflector (on the surface of this imaginary sphere) can be calculated using GO or GO/GTD. Once the fields over this small sphere are known, we may expand them in terms of a spherical wave series on the surface of this sphere and obtain the spherical wave coefficients as before, by orthogonality. Once the spherical wave expansion of the subreflector scatter pattern is known, it can be used to calculate the fields incident on the main reflector.

Spherical wave theory generally yields results which are much more accurate than GO/GTD, but also requires more computation time. For many applications it is excessive, but in the case of very low-noise antenna systems, its use may be necessary.

REFERENCES

1. Rusch, W.V.T., "Scattering from a Hyperboloidal Reflector in a Cassegrainian Feed System," *IEEE Trans. Antennas Propagat.*, pp. 414–421, July 1963.
2. Pogorzelski, R.J., "A New Integration Algorithm and its Application to the Analysis of Symmetrical Cassegrain Microwave Antennas," *IEEE Trans. Antennas Propagat.*, Vol. AP-31, pp. 748–755, September 1983.
3. Ludwig, A.C., "Computation of Radiation Patterns Involving Numerical Double Integration," *IEEE Trans. Antennas Propagat.*, Vol. AP-16, pp. 767–769, November 1968.
4. Rusch, W.V.T. and O. Sorenson, "The Geometrical Theory of Diffraction for Axially Symmetric Reflectors," *IEEE Trans. Antennas Propagat.*, pp. 414–419, May 1975.
5. Lee, S.W., *et al.*, "Diffraction by an Arbitrary Subreflector: GTD Solution," *IEEE Trans. Antennas Propagat.*, Vol. AP-27, pp. 305–316, May 1979.
6. Pathak, P.H. and R.G. Kouyoumjian, "The Dyadic Diffraction Coefficient for a Perfectly Conducting Wedge," *Ohio State University Technical Report 2183-4,* June 5, 1970.
7. Kouyoumjian, R.G. and P.H. Pathak, "A Uniform Geometrical Theory of Diffraction for an Edge in a Perfectly Conducting Surface," *IEEE Proc.*, Vol. 62, pp. 1448–1461, November 1974.
8. Potter, P.D., "Application of Spherical Wave Theory to Cassegrainian-Fed Paraboloids," *IEEE Trans. Antennas Propagat.*, Vol. AP-15, pp. 727–736, November 1967.
9. Wood, P.J., "Field Correlation Diffraction Theory of the Symmetrical Cassegrainian Antenna," *IEEE Trans. Antennas Propagat.*, Vol. AP-19, pp. 191–197, March 1971.
10. Ludwig, A.C., "Near-Field Far-Field Transformations Using Spherical Wave Expansions," *IEEE Trans. Antennas Propagat.*, Vol. AP-19, pp. 214–220, March 1971.
11. Cha, A.G., "Physical Optics Analysis of a Four-Reflector Antenna," *IEEE Trans. Antennas Propagat.*, Vol. AP-34, pp. 992–996, August 1986.
12. Wood, P.J., *Reflector Antenna Analysis and Design,* Peter Peregrinus, 1988.

Index